Exploring geology on the Isle of Arran

The Isle of Arran has an amazingly rich and varied geological history spanning nearly 600 million years, and it has now become a classic location in which to begin exploring geology in the field. This workbook teaches the practical field skills needed by anyone, A-level student, undergraduate, or just keen amateur, who wants to find out more about the rocks that shape our landscape. Step-by-step instructions guide the user through a collection of ten problem-solving exercises based on the most interesting aspects of Arran's geological history. The exercises also introduce concepts that underpin geology as a science and affect the way in which we view the Earth. This approach injects a greater flexibility into the way field geology is taught and shows the subject to be dynamic, thought provoking and fun.

CHRIS NICHOLAS is an experienced field geologist and course leader of students at all levels, and has spent many years exploring and teaching geology, particularly on Arran. He was recently involved in the production of the major BBC TV series *Earth Story* and is currently at the Department of Earth Sciences, University of Oxford.

Exploring geology on the Isle of Arran

A set of field exercises that introduce the practical skills of geological science

C. J. Nicholas

CAMBRIDGE UNIVERSITY PRESS
Cambridge, New York, Melbourne, Madrid, Cape Town, Singapore,
São Paulo, Delhi, Dubai, Tokyo, Mexico City

Cambridge University Press
The Edinburgh Building, Cambridge CB2 8RU, UK

Published in the United States of America by Cambridge University Press, New York

www.cambridge.org
Information on this title: www.cambridge.org/9780521635554

First published 2000

A catalogue record for this publication is available from the British Library

Library of Congress Cataloguing in Publication data
Nicholas, C. J. (Christopher John), 1968–
Exploring geology on the Isle of Arran / C.J. Nicholas.
p. cm.
Includes index.
ISBN 0 521 63555 1 (pb)
1. geology – Field work. 2. Geology – Scotland – Arran, Island of – Guidebooks.
3. Arran, Island of (Scotland) – Guidebooks.
1. Title
QE45.N53 1999
554. 14'23–dc21 98–21345 CIP

ISBN 978-0-521-63555-4 Paperback

'Let us turn from the lessons of the lecture-room to the lessons of the crags and ravines, appealing constantly to nature for the explanation and verification of what is taught. And thus, whatsoever may be your career in the future, you will in the meantime cultivate habits of observation and communion with the free fresh world around you – habits which will give a zest to every journey, which will enable you to add to the sum of human knowledge, and which will assuredly make you wiser and better'

Sir Archibald Geikie, 1871

Director General of the Geological Survey of Great Britain

Contents

Preface

It was during the summer of 1787 that James Hutton and his colleague, John Clerk, succeeded in making the first serious geological survey of the Island of Arran. Since that time many have followed in their footsteps and the ferries now carry geological field parties over to Arran almost every month of the year. The reason for this popularity is that the island has a rich and varied geological history spanning nearly 600 million years packed into its relatively small geographic area. This makes it the ideal training ground in which to put geological science into practice.

Geology as a science is now over two hundred years old and is undergoing a major transformation with the dawn of the twenty-first century. Technological advances are being made at an increasing rate and are providing a multitude of new ways in which to investigate the Earth's dynamics. Consequently, geology is now amongst a number of disciplines collected together as the 'Earth Sciences'. With such a diversity of approaches it is more important than ever for a student of the Earth Sciences to have a solid, central, supporting knowledge of geological processes and principles from which to branch out into further, more specialised studies. To master geology in the field is still the most intensive and effective way in which to achieve this as it develops the mind to think quickly in a logical and scientific manner, as well as introducing practical methods of dealing with rocks on a large scale.

Many geological field trips follow the traditional format of walking a group of students over the rocks during a day in the field, pointing out localities of interest and posing questions. Two published excursion guides currently exist which follow this approach and ably summarise much that is of geological importance on the Isle of Arran. However, those who have led field trips in this way have often found that the problem with this method of teaching is that students must all walk at the same pace and think at the same speed. Everyone, of course, is different; some walk more slowly and arrive as the main group is about to leave the outcrop, whilst others may need a longer look but find time is too short. Consequently, this regime can be frustrating to both students and leaders alike.

This book is designed to inject a greater flexibility into the way geology can be explored in the field; not necessarily by replacing day excursions entirely but by supplementing them with a different approach. It provides a collection of ten, observation-based, structured exercises graded according to their difficulty and covering the spectrum of *'field skills'* that a novice student in field geology needs to master. The work can be carried out in pairs or small groups within the confines of each exercise area with data and observations being recorded directly onto pages within the book. This freedom allows everyone to work at their own pace and encourages solution of geological problems individually or through discussion with friends. Leaders can attend to those who need more help whilst the others have a relatively free rein to explore the subject. The structure of the exercises and the provision for answers also makes it easy to mark and assess a student's progress.

Exercises can be attempted in any order. Alternatively, the book can be treated as a field course in itself with exercises becoming progressively harder over the days and drawing on the skills learnt in earlier ones. Following this philosophy, there are two principal aims of the book:

1 To teach the basic skills of field geology by guiding students through a series of field-based problem-solving exercises.
2 To demonstrate some of the more interesting aspects of Arran's geological history by setting the exercises around them.

A field geologist uses a whole host of field skills in order to gain the maximum amount of information from the rocks. These might range from simply using a hand lens effectively to constructing a geological map with cross-sections. The practicalities of these skills are rarely found explained in text books and are still mostly taught by word of mouth. So it is a further aim of this book to 'demystify' field geology by listing what might be thought of as the twenty-five most commonly used skills. Each one has been allocated a number and is explained as simply as possible with notes and diagrams in Part 4 of this book. Whenever a field exercise requires the use of a certain skill, the corresponding number is given so that it can be referred to if necessary in Part 4. It is hoped that gradually the user will turn to this section less and less as the skills become second nature. However, practice makes perfect, and whilst practising it is extremely useful to have the technique written down somewhere to check on if in trouble.

Many of the principal ideas behind the science of geology date back to the nineteenth century or earlier. So several exercises in Part 3 include a historical slant specifically to introduce some of these and how they were accepted by the scientific community of the time. This also helps to show how science actually 'works' as well as providing a glimpse of some of the more influential early geologists and how their theories still affect the way in which we view the planet today. Above all, this book is intended to show that field geology is dynamic, thought provoking, fun and as crucial in geological teaching today as it has always been.

C. J. Nicholas

Acknowledgements

The first geological field course that I ever went on was to the Isle of Arran. At that time I could not have predicted that I would return for many years and eventually write a book. It began in earnest around 1991 with a few simple field exercises for the undergraduates at the Department of Earth Sciences in Cambridge. From that time I have been fortunate to have the encouragement of many people from universities around the country who also lead field trips to Arran year after year whatever the weather. In particular at Cambridge I would like to thank Andy Buckley. Many of us who thought we knew a thing or two about field geology have passed before Andy's perceptive eye and learnt more than we might like to admit in public. But others who have always taken a keen interest in how to teach field geology, and the fortunes of these exercises, are Paul Pearson, Pete Ditchfield, Mark Hallworth and Simon Price. Once the book was truly underway, additional volunteer help in the field was provided by Liz Hide, Steph Lewis and Clare Glover. Also, my thanks to Rose Edwards for her continual support and patience during the final stages of writing and editing. Finally, I am endebted to all those who have helped at Cambridge University Press. All topographic maps in this book are based on the 1:25 000 Ordnance Survey Outdoor Leisure Map 37 for the Isle of Arran (1995).

C. J. Nicholas
March 1998

Acknowledgements

The first geological field course that I ever went on was to the Isle of A[illegible]. At the time I could not have predicted that I [illegible] for many years [illegible] field courses [illegible] at the Department of Earth Sciences [illegible] that time I have been fortunate to have [illegible] Cambridge [illegible] taken [illegible] in how to teach [illegible] McKellworth [illegible] since the book [illegible] [illegible] Also my thanks [illegible] [illegible] Survey [illegible] Isle of Arran [illegible]

[illegible]

Part 1

An introduction to the field work

1 About the exercises

The exercises in this book are based as far as possible on the geology that can be seen in the field in order to keep specialised knowledge to a minimum. Because of this it is intended that they can be attempted by any enthusiast with an elementary knowledge of geology. Typically the geological content in the field exercises is at a level that would cover A-level to first/second year single or combined honours undergraduate courses.

WHAT YOU ALREADY NEED TO KNOW

To keep this book small enough to be useable in the field it is not possible to provide background information on all the topics touched upon during the exercises. Indeed, its real purpose is to put into practice what can be learnt indoors. So although specific ideas involved in each exercise are explained before field work commences, it must be assumed that the user of this book has at least a very basic prior knowledge of geological processes and principles. However, below is a brief guide to what this knowledge should perhaps include.

1 An awareness of relative and absolute geological time scales.
2 A grasp of the concepts behind plate tectonics and the internal structure of the Earth.
3 An idea of where magma originates and how it might produce intrusive and extrusive igneous rocks.
4 An awareness of basic structures such as faults, folds and how they might form.
5 Some idea of the processes at work in modern-day marine, fluvial, glacial and aeolian sedimentary environments.
6 Some experience of identifying invertebrate fossil groups and interpreting their mode of life.
7 An awareness of at least some of the more common rock-forming minerals in hand specimen. Ideally this should include quartz, feldspars (alkali and plagioclase), micas (biotite and muscovite), pyroxenes, amphiboles, olivine and calcite.
8 Some experience of outdoor activities and basic skills such as how to take and read grid references, how to navigate from a topographic map and how a magnetic compass works.

THE STAR RATING SYSTEM (*)

Each field exercise has been allocated a star, or number of stars. This is intended to show the user at a glance how difficult the exercise is likely to be. The 'difficulty' can be thought of as a combination of three factors:

1 The complexity of the geological concept(s) behind the exercise.
2 The subtlety and detailed nature of the field observation and interpretation needed.
3 The overall time and effort that an exercise requires to complete.

The exercises become more complex in a practical sense from numbers 1 to 5, requiring additional and more varied skills in the *collection* of field data. From exercises 6 to 10, the emphasis changes to the *interpretation* of field data and this increases the intellectual demand. Therefore, the stars allocated to exercises reflect this relative measure of difficulty within the confines of the book.

EXERCISE FORMAT

Ideally geological field work should be a logical, thorough and objective exercise in data collection, concluding with an interpretation of the evidence. Typically the data will be in the form of observational evidence or structural measurements and so maintaining impartiality can be a problem. However, when conducted effectively, field work still maintains a central role in the geological sciences of today and can often be the only means of understanding data generated later in the laboratory.

Learning how to order thoughts and ask the right questions in the field is difficult and is generally taught by 'trial and error'. The problem is addressed in this book by each exercise setting a specific aim or '*Task*'. Subsequent instructions are geared towards what needs to be investigated in order to solve the problem or achieve that task. Consequently, the layout of all exercises is designed to show the beginner how to approach field work in a number of different situations, what to look for and what sort of questions they should be using the rocks to answer before drawing a conclusion from the work. Below is a guide to this format; all exercises are arranged in the same way.

1 Title (*exercise name*). Each exercise has a title followed by its approximate location on the island in brackets. Titles for the first five provide an immediate idea of the broad subject of the geological sciences involved in the field work. The last five exercises have headings that reflect the theme behind each.

2 Task (*what is the point of the exercise?*). This is the aim of the exercise and is stated as briefly as possible.

3 Logistics (*where is it set and how do I get there?*). Each exercise has a '*Start Point*'; in all cases this is either a car park or lay-by on the main road that circumnavigates the island (A841) or the central road known as the 'String' (B880). Also, these are all at or within walking distance of a bus stop. Instructions about how to get onto the exercise area and find the first locality begin from this point and it is a useful rendezvous at the end of the day. The remainder of the comments in this section describe how to find the exercise area on the map and where the nearest village is, and then finally there are directions off the road to begin work.

4 Length (*how long will it take?*). This gives a total recommended time to be spent on the exercise, including getting from and then returning to the '*Start Point*'. There is also a breakdown of the time spent walking to the first locality and then back at the end.

5 Field skills required (*what will I have to use to do it?*). The field skills taught for the first time or re-used in the exercise are given in square brackets. Each is numbered in order of its appearance in the book and can be found described in Part 4. Skills that are being met for the first time are in bold type, whereas those previously introduced are written in plain type.

6 Background information (*the theme, or, why do the exercise?*). Any further information that may be useful to know before beginning the field work is laid out in this section. For the first five exercises (Part 2) this consists of relatively brief comments on the geological processes and principles to be encountered. However, this section in exercises of Part 3 will explain the background to the task in hand, often expanding on the theories or concepts explored during the field work, and so it is vital to read this before proceeding. This section should also help any decisions about choice of exercise if they are being attempted in a different order from that given.

7 The Field Exercise (*how do I go about the field work?*). This is the main part of each exercise and contains full instructions for the field work. It begins by listing a series of steps that, when followed, should lead to a solution of the problem or achieve the task set.

The instructions then continue by leading the user through each major step in field work at a time. Included in these steps are questions that should be answered using field observation and interpretation. The answers should be written in the blank field note pages provided at the back of each exercise before moving on to the next question. Any work-sheets or base maps needed are included in this section and should be annotated as required.

In order to keep the exercise instructions simple and uncluttered, practical field skills and geological terms are not explained or expanded upon in this section unless they are particularly critical to the understanding of the concepts behind the exercise. Those familiar with them can proceed about the work without delay, whilst any who are unfamiliar with the techniques or terms will find them explained at the back of the book in either Part 4 or the Glossary respectively.

8 Help! (*which localities or features should be seen during the field work?*). This is not an *answers* section as such, but it does point the user of this book in the right direction if the clouds of confusion have well and truly descended. The contents of this section will vary depending on the exercise. For instance, it may be photographs and locations of vital outcrops that provide key evidence, or descriptions of features that, when considered together, will point towards a particular solution to a task. It may be that after consulting the **'Help !'** section some localities have to be revisited.

'Help!' is only provided for the more complicated exercises (those with *** or more), but solutions to most practical problems for any of the ten exercises can be found in Part 4. It is worth remembering that this deals with a variety of geological features that might be encountered on Arran. So if problems are being encountered in interpretation of some structure for instance, it may be that it is described in Part 4 somewhere.

To print a full set of answers to each exercise alongside the questions would destroy one of the primary aims of this book, which is that the users discover the geological field evidence for themselves. But the book is deliberately constructed so that if the text is read carefully, the rocks are examined in detail and the field skills are referred to where indicated; then hopefully the user will be guided towards a geologically acceptable solution. Ultimately, this will depend on what has been seen and what is made of it. Therefore, the philosophy behind this book is not for the user to search in vain for 'the right answer', but for them to learn how to reach a scientific solution that others will accept. Finally, if sleep is still being lost weeks after returning from the Isle, then the author can be contacted c/o the Editor (Physical Sciences), Publishing Division, Cambridge University Press, the Edinburgh Building, Shaftesbury Road, Cambridge, CB2 2RU.

2 Equipment

All exercises require the same basic equipment outlined below. It is best to go fully equipped to Arran, but there are shops in Brodick where most of this can be bought if vital items are forgotten or lost during field work.

ESSENTIAL OUTDOOR EQUIPMENT (*You will need to wear or carry these items at all times*)

1 Walking boots with non-slip 'Vibram' soles. Wellington boots can be used ('Hunter' wellies are particularly good); but be aware that they can slip easily on wet rocks.
2 Lightweight trousers to walk in; 'tracksuit' trousers are fine but not jeans; they hasten the onset of hypothermia when wet.
3. Waterproof jacket and trousers; these days 'Goretex' fabrics are popular, but they are expensive and jagged rocks have no respect for how much your clothes cost.
4. Spare jumper, gloves and hat. Synthetic 'fleece' jackets, etc., are particularly warm but are generally not wind-proof.
5 Small first aid kit.
6 Emergency polythene survival bag, whistle and torch.
7 High energy food (chocolate bars!) and drink (at least 1 litre, preferably with sugar in it).
8 1:25 000 Ordnance Survey Leisure map, Isle of Arran (yellow cover). Whilst the maps drawn in this book are as accurate as possible, they only show relatively small areas and it is easy to walk off them. Therefore, it is vital to carry this large-scale map of the whole island at all times.

ESSENTIAL GEOLOGICAL EQUIPMENT (*You will need the first four items for all exercises*)

1 Hand lens (10× or 15× magnification is best).
2 Compass–clinometer (Silva or Suunto models are student-priced). If you cannot afford one, you can buy a compass quite cheaply and make your own clinometer (see Part 4, field skill [10]).
3 Pencils (a propelling 0.5 mm HB and a range of coloured pencils) and drawing pens (Pilot or Edding disposable 0.1 mm pigment ink pens are cheap and excellent).

4 Penknife for hardness tests (any small blade will do, or you can improvise with the side of a metal hand lens).

Optional extras:

5 Calculator.

6 Hard hat (advisable if working near any cliffs or overhangs).

7 A geological hammer may be of occasional use, but the sad truth is that a freshly hammered surface can now be found on almost every outcrop on Arran so please do not add to this erosion.

8 Small bottle of 10% hydrochloric acid (as a test for the presence of calcium carbonate).

9 Small (5 m) retractable metal tape measure.

10 Camera and colour print film (200 ASA or faster).

11 1:50 000 Geological map of Arran (1987) (available from British Geological Survey, Keyworth, Nottinghamshire, NG12 5GG).

12 British Museum (Natural History) Handbook, *British Palaeozoic Fossils.*

3 Safety

Personal safety and that of other members in a group is of paramount importance in field work. No exercise in this book is located in a particularly dangerous or isolated area, yet accidents can and will still happen if appropriate precautions are not taken. For instance, do not go into the field alone. This book can be used perfectly well in a pair or a small group and if anything it is more rewarding to discuss field observations and their interpretation with others at the time that they are made.

Each member of a field party should be aware of basic mountain skills and safety before going to Arran. These are laid out clearly in the booklet *Safety on Mountains: An Approach to Mountain Adventure for Beginners* (1988), published by the British Mountaineering Council, Crawford House, Precinct Centre, Booth Street East, Manchester, M13 9RZ. In addition, anyone taking a group of students to Arran as a field course should obtain and comply with the Committee of Heads of University Geoscience Departments (CHUGD) pamphlet *Safety in Geoscience Fieldwork: Precautions, Procedures and Documentation* (1994), available from the Geological Society, Burlington House, Piccadilly, London, W1V OJU.

There is not sufficient space here to discuss all of the points raised in these two publications, but those principal hazards which might be encountered during geological field work on Arran are outlined below (these are drawn from the CHUGD pamphlet).

COMMON OUTDOOR HAZARDS

1 Lack of adequate equipment. Always carry a compass, map, simple first aid kit and survival bag. Wear proper boots and *not* shoes.
2 Lack of adequate clothing. The weather on Arran at any time of year is variable to say the least. The temperature can drop sharply if you walk into cloud, or if the wind picks up. Getting wet will have the same effect, so wet and windy conditions are the fastest way to get hypothermia. Carry some spare, dry clothing in case.
3 Isolation. Never go into the field alone. Common injuries such as a sprained ankle could easily require help to get down to the main road or the nearest house. It has already been stated that the exercises in this book should be attempted in pairs. Even so, it is still important to make sure that someone knows the days' itinerary and the expected time of return.
4 Illness and allergies. It is important to be aware that some illnesses, allergies or disabilities may prove hazardous in potentially cold mountain terrain. Ask your doctor if you are in any doubt.

POSSIBLE HAZARDS IN FIELD WORK ON ARRAN (*With exercises where this hazard is most likely to be encountered*)

1 Steep, unstable or slippery rocks such as cliffs (exercise10), screes (exercises 6, 7, 10) and intertidal areas with seaweed-covered boulders (exercises 3, 4, 6, 8, 10).
2 Falling rocks. Be particularly careful under any overhangs or near cliffs. A hard hat should be worn in any such areas (exercises 6, 10).
3 Disused mines, quarries and caves. Never enter any of these under any circumstances (exercises 5, 6).
4 Rapid changes in the local weather, which are particularly common in the upland areas on Arran and may include sudden loss of visibility and falling temperatures (exercises 1, 5, 7, 9).
5 Hypothermia. This can be from inadequate clothing, getting wet, high winds, exhaustion, or a combination of these factors. It could apply to any exercise, but particularly the longer ones (exercises 9 and 10).
6 Bogs. They often look like solid ground but can be over 1 m deep. Always test the ground ahead before shifting your body weight forward in a boggy area (exercises 1, 5, 9).

7 Mountain streams. They have more energy than you think. Always use a bridge to cross them. A glance at the large boulders in the stream bed should demonstrate the potential energy of a mountain river in flood (exercises 1, 5, 7, 9).
8 Tides. Purchase or check a tide timetable (available from bookshops/ newsagents on Arran; for instance 'Brodick Books' near the ferry terminal). Always be aware of the state of the tide and avoid isolated promontories and headlands when it is coming in (exercises 3, 4, and *especially* 8).
9 Waves. Stormy weather can cause freak waves, so avoid standing right at the waters' edge on coastal sections (exercises 2, 3, 4, 6, 8, 10).

Finally, it is up to the user of this book to decide whether a situation is potentially dangerous. The author accepts no liability for any injury incurred whilst in pursuit of field work using this book.

4 Logistics and itinerary

The Isle of Arran is a fifty-five minute ferry journey from Ardrossan harbour on the mainland Ayrshire coast. There are five regular daily sailings Monday to Saturday (but only four on Sundays) and the ferry can carry coaches as well as minibuses and cars (vehicles should be booked in advance with Caledonian MacBrayne Ltd, The Ferry Terminal, Gourock PA19 1QP). Foot passenger tickets can be bought at the harbour terminal before boarding. Ardrossan harbour is served by a rail-link to Glasgow Central Station (the journey time is just over an hour).

The ferry arrives at Brodick on the east coast of Arran where there is also a bus station. The local bus routes regularly serve all of the villages closest to the exercise areas in this book such as Corrie, Sannox, Lochranza, Catacol, Pirnmill, Machrie and Blackwaterfoot. Details of where to stay on Arran and a bus time-table can be obtained from the tourist information centre located at the Brodick ferry terminal (Isle of Arran Tourist Board, Tourist Information Office, Brodick, Isle of Arran KA27 8AU). At the time of going to press there were three main bus routes on the island: one around the north of the island, one around the south and one over the String road in the centre. These are supplemented by the post bus (which can carry some passengers).

Figure (i) shows the simplified geology of Arran and the location of each field exercise area with its proximity to the nearest village and bus route. For the age and stratigraphic relationship of each unit see Figure (ii). Exercises in this book are arranged so that all ten can be attempted in sequence over the course of about seven days. The timings for each exercise are only suggested and need to be flexible if public transport is used on the island. However, if two are attempted in order each day until exercise 9 then the morning and afternoon's work should complement or contrast to give some variety to the day. These pairs are also arranged to keep a minimum of journey time between them. Below is a suggested itinerary for attempting all ten exercises during a week on Arran using public transport.

NIS 324 = North Island bus service, SIS 323 = South Island bus service, S322 = String road bus service. In addition, all exercise areas can be reached using the Post Bus.

Day 1 *NIS 324 to Sannox*
am: 1. Present day processes *
walk or NIS 324 Sannox to Corrie
pm: 2. Clastic sedimentary rocks *
Day 2 *NIS 324/S322 to Machrie*
am: 3. Intrusive igneous rocks *
NIS 324 Machrie to Catacol
pm: 4. Changing the record: metamorphism and deformation **
Day 3 *NIS 324 to Sannox*
am: 5. 3-D Thinking in time and space **
walk or NIS 324 Sannox to Corrie
pm: 6. Cuvier's catastrophies ***
Day 4 *NIS 324 / SIS 323 / S322 to Heritage Museum*
am: 7. Prof. Speyside's theory ***
NIS 324 Heritage Museum to Catacol
pm: 8. The curious case of 'Catacol cairn' ***
Day 5 *NIS 324 to Sannox*
am and pm: 9. Dr Hutton's dilemma ****
Day 6 *NIS 324 to Sannox*
am: 9. Dr Hutton's dilemma ****
NIS 324 Sannox to Blackwaterfoot
pm: 10. Pluto's revenge ****
Day 7 *NIS 324 / SIS 323 / S322 to Blackwaterfoot*
am and pm: 10. Pluto's revenge ****

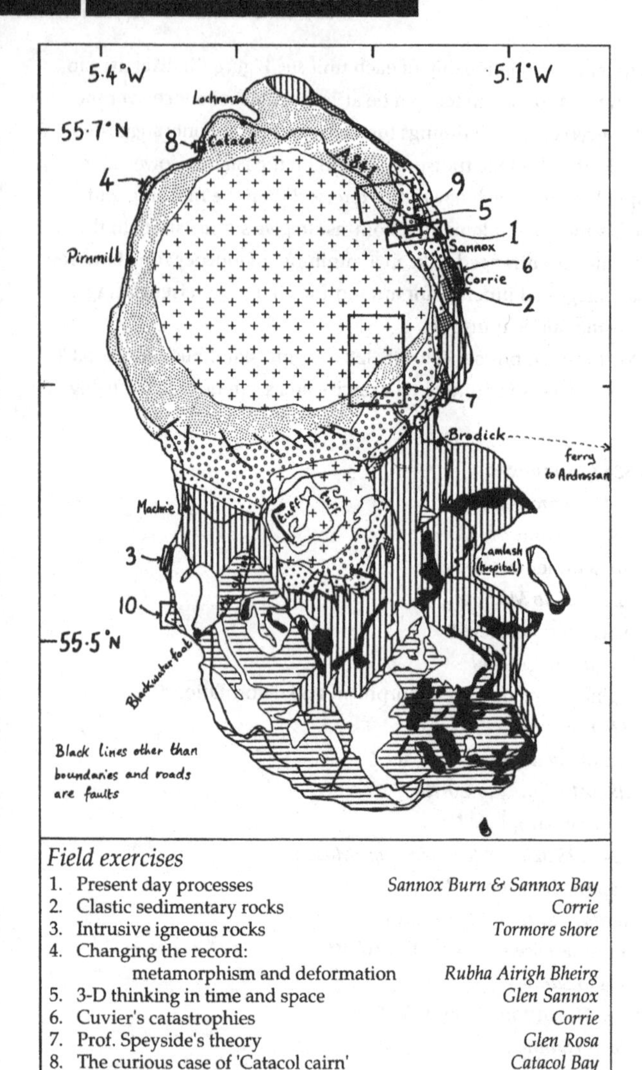

FIGURE (i) Simplified geological map of Arran showing the location of each exercise area. The shading used is the same as for Figure (ii), where the name and age of each geological unit can be found. Both are based on the 1:50 000 Solid Edition (Special District) geological map for the Isle of Arran issued by the British Geological Survey (Scotland) 1987.

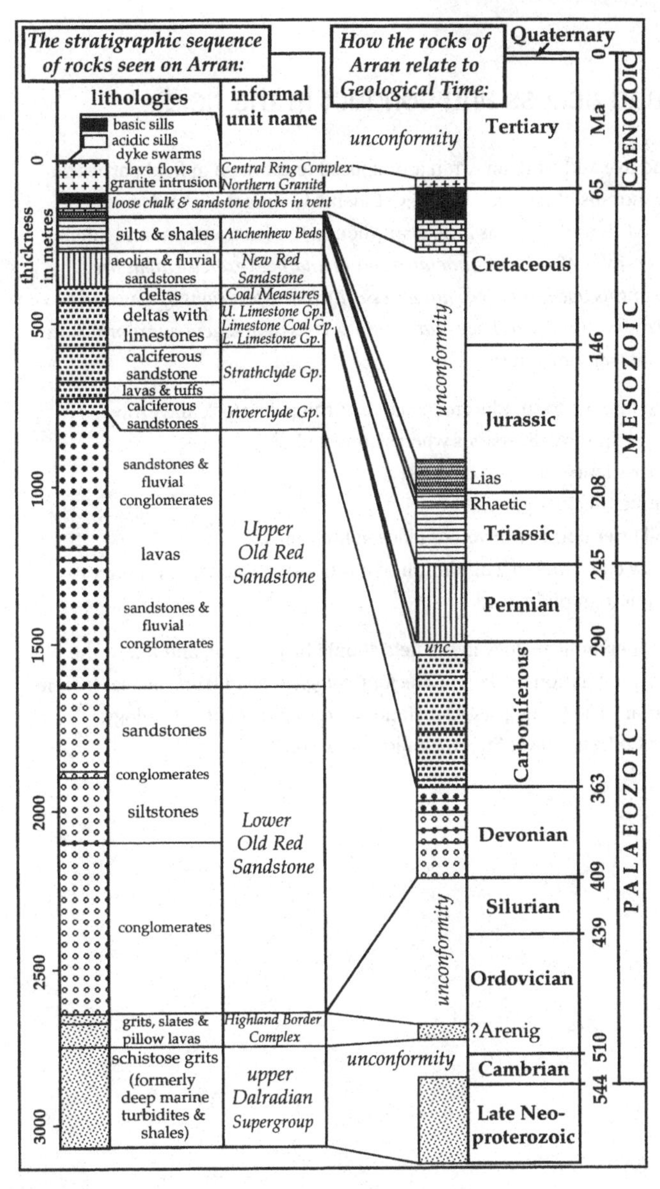

FIGURE (ii) Stratigraphic sequence and age of the principal geological units on the Isle of Arran.

5 Land access and conduct in the field

Those who live and work on Arran are exposed to a stream of field parties during many months of the year. Geology students visiting the island should be sensitive to this and cause as little disruption to the environment and inhabitants as possible. *Whilst recommended routes and field exercise areas are described in this book, this does not necessarily indicate a right of way and it is the user's responsibility to check and obtain permisssion where appropriate.* In particular, it is important to:

1 Keep to marked footpaths and avoid disturbing livestock and crops.
2 Follow temporary diversions when signposted.
3 Only cross fences at stiles or gates.
4 Remember to close gates.
5 Take all litter home with you or place in litter bins.
6 Keep any collecting to a minimum (if it is necessary at all); it is better to take a photograph instead.

Anyone persuing geology in the field should be aware of, and follow the Geologists' Association leaflet; *A Code for Geological Field Work*, available from The Librarian, The Geologists' Association, c/o Department of Geology, University College, Gower Street, London WC1E 6BT.

Part 2

Learning basic field skills

1 Present day processes *

(Sannox Burn and Sannox Bay, north-east coast)

PRELIMINARIES

Task To investigate weathering and erosion of mountains and find out what happens to the transported products.

Logistics *Start Point*: Sannox Bay car park and telephone box

[NS 0161 4540]

Glen Sannox and Sannox Bay are on the north-east coast of Arran. The nearest village to the exercise area is South Sannox [NS 018 452]. Access to the exercise area is via the signposted cart track into Glen Sannox which can be found on the opposite side of the road from the *Start Point.*

NB This exercise requires studying wet sediment by putting it on diagrams in this book so you need to have some piece of transparent plastic to place over the page to protect it. A thin fragment of polythene bag (that perhaps had your lunch in) is ideal. You will also need to buy an orange (!) if you wish to perform the velocity experiment on Sannox Burn.

Length

3 hours: 25 min walk to begin first step, 2.5 hr to complete the exercise, 5 min walk back to *Start Point.*

Field skills required [1] [2] [3] [4] [5] [6] [7]

Background information

Arran's northern mountains are the focus of this first exercise which involves following a river's course from upland area to the sea (Figure 1.1). It introduces the fundamental geological concepts of continual land erosion as part of a 'rock cycle' and that present day processes provide the basis upon which to interpret those of the past (the second half of the rock cycle, that of burial, deformation and uplift, is explored in Exercise 4). This exercise begins with the determination of whether the rocks forming the mountains are of an igneous, sedimentary or metamorphic origin. It then continues by examining how these mountains are weathered, eroded and transported as river sediment and it introduces a mixture of current clastic sedimentary processes at work in fluvial and coastal environments. Finally, these observations of modern day processes are used to predict the potential rock record.

FIGURE 1.1 Looking west into Glen Sannox and the heart of the northern mountains.

THE FIELD EXERCISE

Read through all the instructions thoroughly before beginning work. Annotate maps and figures as required and write your answers to the questions as field notes on the pages provided at the end of the exericse.

Steps needed to achieve the set task

1 Find an example of weathered material in a mountainous area.
2 Identify the bedrock.
3 Establish how the bedrock is weathering.
4 Sample river sediment adjacent to bedrock weathering.
5 Sample river sediment during transportation to the sea.
6 Sample river sediment at confluence with the sea.
7 Synthesis.

Find an example of weathered material in a mountainous area

It would be easy to spend half a day or so wandering around Glen Sannox looking for good examples of weathering and erosion in the mountains. To save time though a suitable starting location is shown in Figure 1.2; circled '1' on the map at [NR 9964 4501]. It is situated on the north bank of Sannox Burn about 2 km up the valley from the *Start Point*. To find it will require field skill [1], combining map reading with observation of key landmarks in the valley.

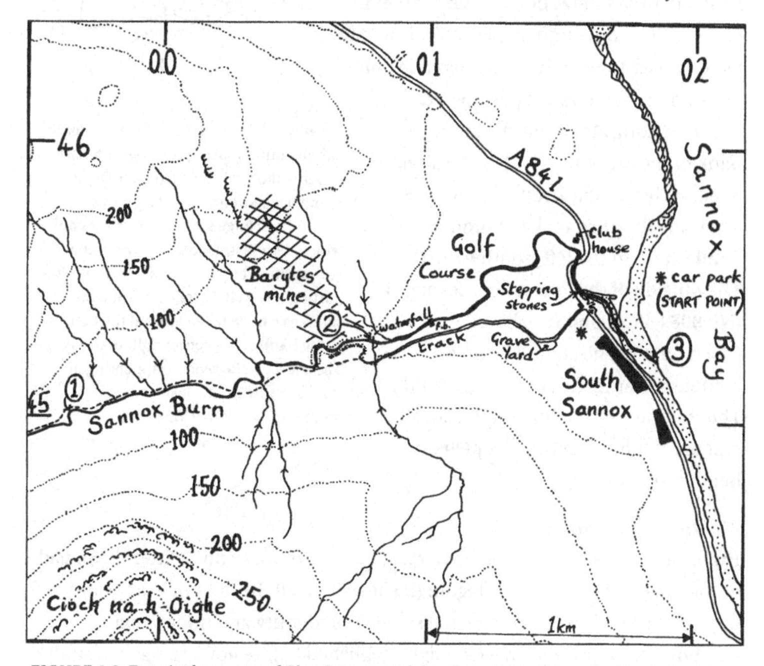

FIGURE 1.2 Exercise base map of Glen Sannox, with localities 1, 2 and 3 marked in circles

First, study Figure 1.2 and find the car park at the *Start Point* (on the right hand side) and 'locality 1' to the west (on the left hand side). Then begin walking up the cart track on the opposite side of the road by the cottage, signposted to Glen Sannox. Follow this track into the valley and after about 10 to 15 minutes look out for Sannox Burn on the right hand side. Cross it at the wooden footbridge in a small cluster of trees before the old barytes mine is reached (the bridge is labelled 'f.b.' on Figure 1.2). A path on the northern bank continues on up the valley, passing through the old baryte mine workings (see Exercise 5). There are a number of sheep 'tracks' and it is often difficult to see which path to follow, but they all head in the same general direction; the first test is to navigate successfully to the grid reference point.

Six minor tributaries are crossed whilst walking along this northern bank of the burn; the last four lie along a relatively straight stretch of the river. After this the burn makes a prominent bend round to the left (southwards). The outside of this meander is locality 1 [NR 9964 4501] and the shoulder of Cioch na h-Oighe should lie on a compass bearing of ~164° to the south. The track leaves the burn at this locality and continues off across the peat and heather (Figure 1.3).

FIGURE 1.3
Locality 1 on the outside of a meander in Sannox Burn, looking south–south-east towards the peak of Cioch na h-Oighe in the background. The sketch was drawn from large slabs on the river bank which demonstrate the bedrock mineral composition and its chemical weathering particularly well. Excellent sediment samples can be taken around the boulders at the bank in the centre right of the sketch. The track can be seen disappearing in the far centre right. Scale: boulders in foreground ~1–2 m wide.

Identify the bedrock

Examine the numerous large boulders and slabs of rock which litter the outside of the meander at locality 1. Figure (i) shows that all the 'bedrock' here and further west up the valley to be the same. The identity and mineralogical composition of this rock needs to be established before embarking on a study of weathering, erosion and transport. Hunt around several of the boulders and pick one which shows the three minerals that compose it most clearly.

- Is this rock igneous, sedimentary or metamorphic in origin? **[2] [3]**
- What different types of mineral can you identify in this rock? [2] **[4]**
- What is the approximate size range of each mineral 'species'? [2] [4] (e.g. amphiboles ≈ 1–2 mm, muscovite ≈ 2–4 mm)
- Using Figure 1.4, what percentage of the 'whole rock' does each type of mineral represent? [2]
- What is the name of this rock? **[5]**

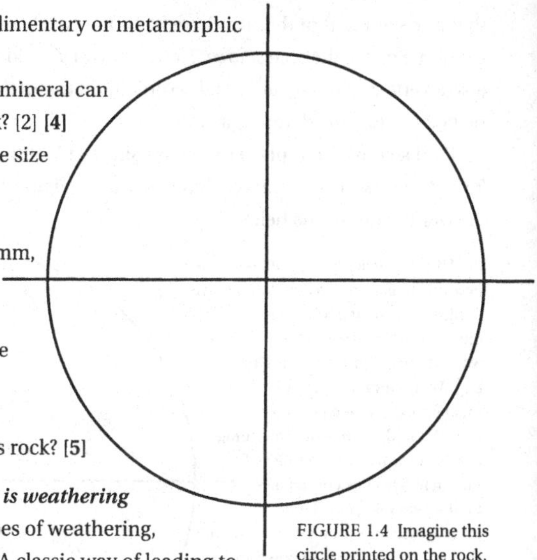

FIGURE 1.4 Imagine this circle printed on the rock, draw in the minerals that you see and estimate the percentage of each mineral present within the circle. For instance, quartz = 60%, plagioclase feldspars = 40%. NB The total must equal 100%!

Establish how the bedrock is weathering

There are two principle types of weathering, mechanical and chemical. A classic way of leading to the first of these is by extremes of temperature which fracture or shatter the rock. Chemical weathering, particularly from weak acids in rainwater or soil, can attack the component minerals in the rock itself. As the minerals are broken down, altered or dissolved their colour can change and this is particularly common in feldspars which turn brown/orange. So the two forms of weathering should cause distinctive changes to the shape and colour of the rocks they attack. It should be clear that the boulders and slabs at locality 1 are undergoing active weathering, producing detritus which is mixing with the acidic, peaty soil.

- Which forms of weathering can be observed at locality 1?
- Is it possible to tell if any mineral is more prone to weathering than the others and if so, which? [2]

Sample river sediment adjacent to bedrock weathering

Rainwater draining down the valley sides washes the weathered rock debris into Sannox Burn. This removal is termed *erosion.* But from the moment of rock disintegration, movement of the products (now qualifying as clastic sediment) begins the process of *transport* and eventually *deposition.* With only one type of bedrock lithology this far up the valley one might expect that any sediment in the river here has come solely from its weathering and erosion, i.e. it is the sole *source*. However, you should have noticed that there is also a veneer of unconsolidated sediment on top of the bedrock and this must be taken into consideration as well.

Find a convenient place on the outside bank of the meander where the 'gritty' sediment is within reach below water (Figure 1.3). Use Figure 1.5 to answer the questions below.

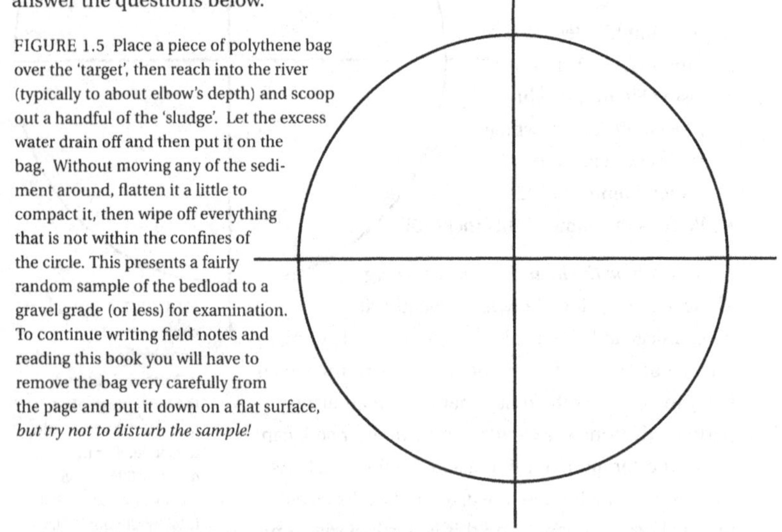

FIGURE 1.5 Place a piece of polythene bag over the 'target', then reach into the river (typically to about elbow's depth) and scoop out a handful of the 'sludge'. Let the excess water drain off and then put it on the bag. Without moving any of the sediment around, flatten it a little to compact it, then wipe off everything that is not within the confines of the circle. This presents a fairly random sample of the bedload to a gravel grade (or less) for examination. To continue writing field notes and reading this book you will have to remove the bag very carefully from the page and put it down on a flat surface, *but try not to disturb the sample!*

- What is the ratio (in %) of individual minerals to fragments of rock (e.g. 30%:70%)? [2]
- What is the size range of the rock fragments (termed 'lithic fragments')? [2]
- What different types of mineral can you identify in the lithic fragments? [2] [4]
- What is the approximate size range of each mineral 'species' in the lithic fragments? (e.g. amphiboles ≈ 1–2 mm, muscovite ≈ 2–4 mm) [2]
- Are all the lithic fragments of the same composition? [2] [4] [5]
- Are the lithic fragments weathered, eroded and transported bedrock?
- Are the loose minerals the same 'species' as those in the lithic fragments? [2] [4]
- Are the loose minerals weathered, eroded and transported bedrock?
- What is the roundness of both the loose minerals and the lithic fragments? [2] **[6]**
- How well sorted is the overall sediment? [2] [6]

- What is the sphericity of the loose minerals and the lithic fragments? [2] [6]
- Approximately what percentage of the whole sludge sample in Figure 1.5 does each component represent (e.g. quartz = 20%, granite fragments 30%; make sure to include both the lithic fragments and loose minerals)? [2]
- Are there any clasts that have not come from the source rock, and if so, where might they have come from? [2]

Sample river sediment during transportation to the sea

Having documented the initial sediment bedload in the river, any changes in its composition during transport to the sea should be detectable using the same sampling methods at localities further downstream. Any number of places could be picked at which to do this, but try locality 2, which is just above a waterfall on the south bank of Sannox Burn about 1 km downstream (Figure 1.2; [NS 0078 4528]). Figure (i) shows that the bedrock here has changed and a consequent compositional difference in the river sediment might be expected from this new source.

To reach locality 2, walk back down the track on the north side of Sannox Burn towards the sea. There are two options for crossing the burn, but the only

FIGURE 1.6 Locality 2 looking east towards a waterfall marked by the line of boulders across the burn. The path is just out of the photo to the right. Excellent sediment samples can be taken anywhere along the southern bank shown in this photo, a distance of about 5–10 m.

guaranteed dry crossing is to retrace the path all the way back to the footbridge and then return westwards along the footpath on the southern bank. The other alternative is to use a ford on a meander about 200 m upstream from locality 2 (but the depth of water at this ford can be highly variable). Whichever route is taken, locality 2 can be spotted easily from either side of the river because it is at the top of a waterfall where a log stretches across the burn. A second sample of Sannox Burn bedload can be taken on the south bank a few metres to the west of the log (Figure 1.6). The observations that need to be made on this new sample are similar to those at locality 1; however, the emphasis of the investigation here is on how the physical properties of the sediment have changed (if at all) since the last sample more than a kilometre upstream. Use Figure 1.5 to obtain a fresh sludge sample to describe.

- What is the ratio (in %) of individual minerals to fragments of rock (e.g. 30%:70%), and is this any different from locality 1? [2]
- What is the size range of the lithic fragments and is it any different from locality 1? [2]
- Are all the lithic fragments weathered, eroded and transported 'locality 1 bedrock' and if not, what is the composition of the new fragments (what different types of mineral can you identify in them and what is the approximate size range of each mineral 'species')? [2] [4] [5]
- What loose minerals are present in the sediment? [2] [4]
- Are there any new loose minerals and if so, what are they? [2] [4]
- Are there any minerals missing in this sample that were present at locality 1, and if so, what are they? [2] [4]
- What is the roundness of both loose minerals and lithic fragments and how does this compare with locality 1? [2] [6]
- How well sorted is the overall sediment and how does this compare with locality 1? [2] [6]
- What is the sphericity of the loose minerals and lithic fragments and how does this compare with locality 1? [2] [6]
- Approximately what percentage of the whole sludge sample from Figure 1.5 does each component represent (e.g. quartz = 20%, granite fragments 30%; make sure to include both the lithic fragments and loose minerals)? [2]
- Has the sediment composition changed since locality 1 and if so, how might this be explained?

Sample river sediment at confluence with the sea

Other samples could be taken along Sannox Burn to confirm the conclusions of the previous step, but the next major effect on the composition of the river sediment is likely to be at the river mouth. A river's energy falls dramatically when it meets the sea. This decrease means that the river no longer has the power to move much of its coarser bedload. In addition, any clay held in suspension in the river expands on contact with saltwater and sticks together or to other particles until they also become too large to be transported by the water. Both cases result in a noteable increase of deposition where freshwater meets saltwater. Another factor that must be considered at the mouth of Sannox Burn is that a spit has built out from the southerly longshore drift along the coast. Consequently, the bedload sediment deposited here could be a mixture from two separate source areas, one from Sannox Burn and the other from currents moving beach sediment southward in Sannox Bay. This intriguing possibility can be investigated by taking a third and final sediment sample, this time at the tip of the spit; locality 3 (Figure 1.2). To reach it, walk back along the main path to the road and the *Start Point* car park.

If there is time, you might like to try and measure the average velocity of Sannox Burn. This can be done between the stepping stones at the *Start Point* car park [NS 0156 4547] and the roadbridge over Sannox Burn about 100 m to the north. All you need to do this is a watch and an orange. Drop the orange into the river from the roadbridge and record the time to the second that it hits the water. Then run back along the side of the road to the stepping stones and watch the approach of the oncoming orange. Record the time at which the orange reaches the stepping stones. Retrieve it and work out how long it took to cover the distance. Use the equation:

$$\text{velocity (m/s)} = \text{distance (m)} / \text{time (s)}$$

From this you can calculate the average velocity of Sannox Burn in this stretch of the river (the distance from bridge to stepping stones is 120 m).

If time is short or you are unfortunately not equipped with an orange, then continue to locality 3 by crossing the stepping stones. Once on the other side, follow the path through the thorn bushes and onto the beach at Sannox Bay. Walk along the shoreline to the south until the tip of the spit is reached; anywhere just below the waterline here will do for sampling. Use Figure 1.5 again to get a sludge sample on the bag.

- What is the ratio (in %) of individual minerals to fragments of rock (e.g. 30%:70%), and is this any different from the other localities? [2]
- What is the size range of the lithic fragments and is it any different from the other localities? [2]

- Are all the lithic fragments weathered, eroded and transported 'locality 1 bedrock' and if not, what is the composition of the new fragments (what different types of mineral can you identify in them and what is the approximate size range of each mineral 'species')? [2] [4] [5]
- What loose minerals or other clasts are present in the sediment? [2] [4]
- Are there any loose minerals or other clasts that were not present at locality 2 and if so, what are they? [2] [4]
- Are there any clasts that might have their source from longshore drift in Sannox Bay and if so, how is it possible to tell?
- Are there any minerals missing in this sample that were present at locality 1, and if so, what are they? [2] [4]
- What is the roundness of loose minerals, other clasts and lithic fragments and how does this compare with former localities? [2] [6]
- How well sorted is the overall sediment and how does this compare with the other localities? [2] [6]
- What is the sphericity of loose minerals, other clasts and lithic fragments and how does this compare with the other localities? [2] [6]
- Approximately what percentage of the whole sludge sample from Figure 1.5 does each component represent (e.g. quartz = 20%, granite fragments 30%; make sure to include lithic fragments, other clasts and loose minerals)? [2]
- Has the sediment composition changed since locality 2 and if so, how might this be explained?

Synthesis

This exercise has dealt with observing modern day processes at work, but as a geologist this knowledge and experience must be constantly applied to interpret the rock record. In conclusion, consider the sediment sample described at each of the three localities and then imagine it compacted and turned into solid rock ('lithified').

- What would be the name given to the sedimentary rock from each locality? [6] [7]
- If these hypothetical sedimentary rocks were to be found as isolated outcrops on a hillside some time in the future, how would a geologist be able to work out that they were originally deposited in a river?

Exercise 1: field notes exact location	General location: Date:	data

Exercise 1: field notes exact location	General location: Date:	data

Exercise 1: field notes exact location	General location: Date:	data

Exercise 1: field notes exact location	General location: Date:	data

Exercise 1: field notes exact location	General location: Date:	data

Exercise 1: field notes exact location	General location: Date:	data

2 Clastic sedimentary rocks *

(Corrie, east coast)

PRELIMINARIES

Task To examine sedimentary bed-forms in Carboniferous to Permian age rocks and determine their environment of deposition.

Logistics *Start Point*: The Corrie Hotel car park [NS 0255 4310] The village of Corrie is situated on the east coast of Arran. The exercise area is a short section of the beach south of the hotel which has easy access directly from the road. There are intermittent, low-lying outcrops on the foreshore between the hotel car park and a prominent white ridge (which marks the top of the Carboniferous succession) 225 m further to the south [NS 0260 4310]. The investigation begins at a low, eroded bed ~8 m north of this ridge and then proceeds south to the harbour at [NS 0260 4280]. *The exercise is dependent on a low tide during the first two steps*, but once work has progressed onto the prominant white ridge it can be attempted until the tide is high.

Length

2 hours: 5 min walk to begin first step, 1 hr 45 min to complete the exercise, 10 min walk back to *Start Point*

Field skills required [1] [2] [3] [4] [5] [6] [7] **[8] [9] [10] [11]**

Background information

FIGURE 2.1 Looking north along the exercise area from the harbour to the south. The majority of exposure seen here at mid to low tide is Permian in age. The prominent building in the centre right (where skyline meets land) is the Corrie Hotel.

By far the most common sedimentary rocks on Arran are those formed from the disintegration, re-deposition and cementation of other rocks. Such clastic sedimentary rocks outcrop over more than half of the island with Carboniferous to Permian age sediments being particularly well exposed on Corrie foreshore (Figure 2.1). This exercise introduces a variety of forms that sedimentary beds can take in outcrop, such as horizontal, cross-bedded, convolute, rippled, desiccation-cracked or dunes. It examines what this evidence, combined with observations on any fossils found, might tell the geologist about the original depositional environment. The exercise also involves practical experience in using the compass–clinometer for measuring the angle of dip in originally horizontal beds and in cross-sets. The data are plotted on a rose diagram to demonstrate the different dip directions of units, and this is considered in relation to the sedimentary structures present, or to subsequent tectonic tilt of the bedding. The exercise finishes by looking at the potential of some bed-forms to indicate the latitude of Arran in the geological past.

THE FIELD EXERCISE

Read through all the instructions thoroughly before beginning work. Annotate maps and figures as required and write your answers to the questions as field notes on the pages provided at the end of the exericse.

Steps needed to achieve the set task

1 Identify lithology and bedding planes of the first outcrop.

2 Examine bed-forms at the base of the ridge.

3 Investigate the main body of the ridge.
4 Determine the nature of the Carboniferous–Permian boundary.
5 Reconstruct depositional environments of the Permian rocks.
6 Synthesis.

Identify lithology and bedding planes of the first outcrop

To begin the exercise, walk south along the roadside and locate the prominent ridge that runs out to sea perpendicular to the road ~225 m south of the hotel [1] (Figure 2.2). Just a few metres to the north of this ridge is a flat-lying brick-red outcrop about 1–2 m wide that lies below the mid-tide line. It should be searched for amongst the pebbles on a bearing of ~314°, 7–8 m from the north face of the ridge at [NS 0260 4310]. This outcrop is locality 1 (Figure 2.3). At first, examine it by standing back to notice any larger scale features and then take a closer look down on your knees with a hand lens.

- What lithological category does this rock belong to? [2] [3]
- Are any minerals identifiable in this rock and if so, what are they? [2] [4] (In many cases it is useful to note down what is not visible as well as what can be observed)
- What are the physical properties of this rock? [2] [6] (Again, say if you cannot see certain properties)
- What is the name of this rock? [2] [7]

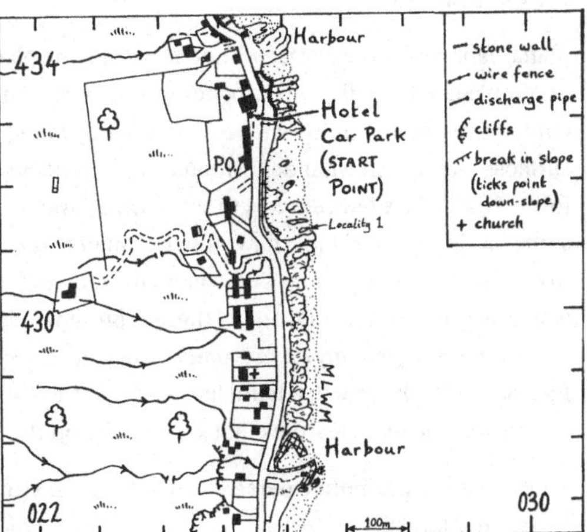

FIGURE 2.2 Exercise base map of Corrie foreshore, with the hotel, locality 1 and the harbour shown.

By following the flow chart in [3], some preliminary observations will already have been made concerning the distinction between bedding, jointing and cleavage at this locality. Performing a hardness test and taking a closer look at the rock with a hand lens should clarifiy whether the

FIGURE 2.3 View north-east from the prominent white ridge. A geologist can be seen hunting for fossils at locality 1.

planar fabric here is cleavage or not. However, the planar surfaces seen in this outcrop could be sedimentary bedding and/or jointing. Luckily, this lithology contains fossils and these can be used not only to indicate what are bedding surfaces, but also in what depositional environment this lithology may have formed. So, take a few minutes to scour what little outcrop there is and find some fossils. Field skill **[8]** shows a selection of fossils commonly found on Arran, so use it to get an idea of what to look for and how to identify the fossils when they are found. Only one of these is present at locality 1. *NB They are not common. If you find any do not remove them; they need to be observed in situ.* Instead, study them with a hand lens and make a few sketches which summarise the main features that you can see on the fossil.

- What is the taxonomic name given to the fossils and to what group do they belong? [2] [8]
- Are the fossils found preserved whole and undisturbed in their 'life position', and if not, then how are they found? [2] [8]
- What is it about the way they have been buried by the sediment that indicates which are bedding surfaces in this outcrop? [2] **[9]**
- What does the fossil evidence suggest as the likely environment of deposition for these sediments? [8] *NB If no fossils have been found then what does the sediment alone suggest about the depositional environment?* [6] [7] [9]
- What is the angle of dip and the dip direction of the bedding (plot the measurements on Figure 2.4)? **[10] [11]**

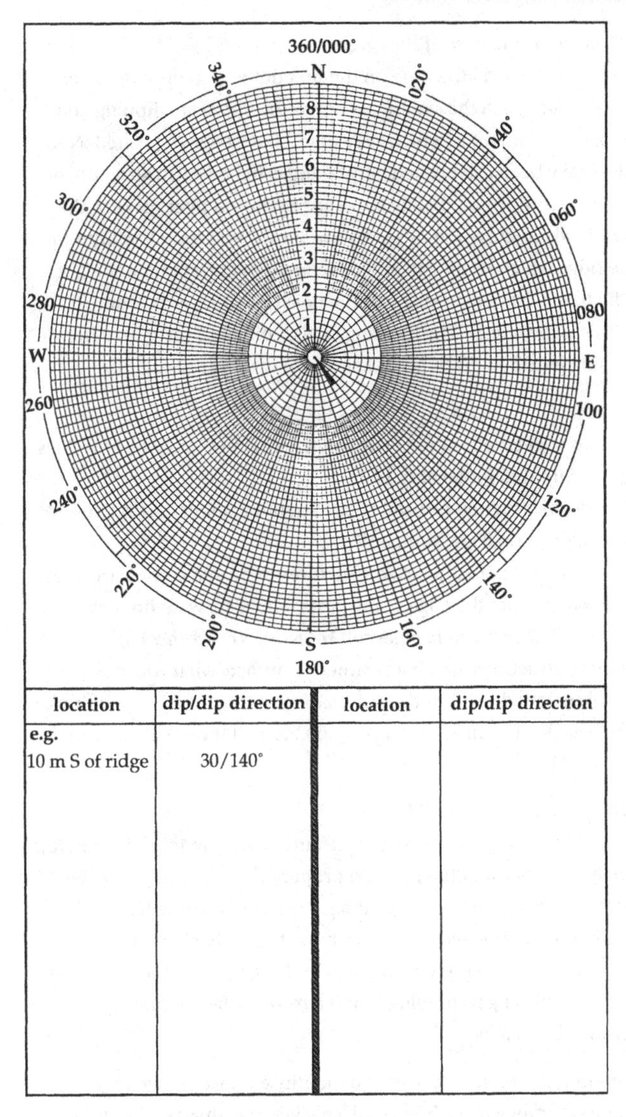

location	dip/dip direction	location	dip/dip direction
e.g. 10 m S of ridge	30/140°		

FIGURE 2.4 Write your location and dip/dip direction measurements in the columns provided. Then, using [11] as a guide, block in the appropriate segment on the rose diagram. An example of how to plot a single value of a dip direction to 140° is shown to help you.Try and plot data from each new locality in a different pencil colour as this will highlight any changes in dip direction between them.

Examine bed-forms at the base of the ridge

It was important to spend a little time on the first outcrop in order to determine the direction in which the sediments of the foreshore are dipping and also the kind of environment in which they might have been deposited. Now that both these have been established, any changes in bed-form structure or lithology should become immediately apparent as the younger rocks to the south are examined. So move the 7–8 m south–south-west from locality 1 to the red/white ridge which should stand between waist and shoulder height. Look along the base of the ridge between the beach pebbles and about 50 cm above them (this is locality 2). The bedding takes a different form here from that at underlying locality 1. Sketch an example of what you see.

- What type of bed-form do the sediments take? [9]
- What does this suggest about the depositional environment and how does this compare with that for locality 1? [9]
- Can the bed-form structure be used to say anything about current flow in the original depositional environment? [9]
- You may be able to see smaller-scale structures in profile at the base of the ridge about 9 m in from the edge of the grass, and if so, what do they represent? [9] (*NB The beach pebbles can sometimes cover these up*)
- Although these structures are on a weathered surface, what was the approximate current direction? (Take a bearing in the direction you think the current was flowing and plot it on Figure 2.4 as if it were a 'dip direction' value) [9] [10] [11]

Investigate the main body of the ridge

The simple fact that the ridge is far more resistant to erosion than the outcrop at locality 1 demonstrates that the physical properties of the ridge must be different. Whether this is due to the way it has been deposited, what has been deposited or what has happened to it since, must be resolved before continuing on towards the Permian strata. So before climbing up onto the top of the ridge, it is worth examining its lithology. Find a fresh surface at about eye-level and study it with a hand lens.

- What different types of mineral can you identify as clasts and what is the approximate size range of each mineral 'species' (e.g. quartz ≈ 1–2 mm, muscovite ≈ 2–4 mm)? [2] [4] [6]
- What is the roundness of each type of clast? [2] [6]
- How well sorted is the overall sediment? [2] [6]

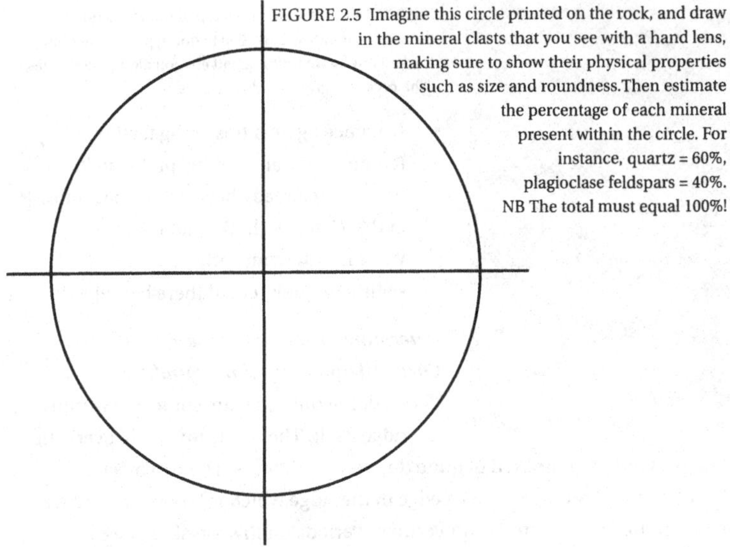

FIGURE 2.5 Imagine this circle printed on the rock, and draw in the mineral clasts that you see with a hand lens, making sure to show their physical properties such as size and roundness.Then estimate the percentage of each mineral present within the circle. For instance, quartz = 60%, plagioclase feldspars = 40%. NB The total must equal 100%!

- What is the sphericity of each type of clast? [2] [6]
- Using Figure 2.5, what percentage of the 'whole rock' does each type of mineral represent? [2]
- What is the name of this rock? [2] [7]
- In what ways have the sediment composition and 'maturity' (sorting, sphericity and roundness) changed since locality 1 and, combined with the bed-form evidence, what might this suggest about the depositional environment for this ridge? [4] [6] [9]
- Is this the same as locality 1?

Now climb onto the top of the ridge and take a few minutes to walk the length of it, examining the form of the sedimentary bedding and perhaps jump back down to locality 2 to see the structures in cross-section (Figure 2.6). Try sketching some particularly good examples of bedding shape and structure (this is locality 3).

- What evidence seen along the side of the ridge suggests that the 'folded' planar surfaces are the original sedimentary bedding?
- Can any dip and dip directions of the bedding be measured, and if so, what are they? (Try and take at least six measurements and plot them on Figure 2.4) [10] [11]
- From Figure 2.4, is there any pattern to these bedding dip directions?

FIGURE 2.6 View along the top of the white ridge, looking east–north-east. The bedding planes are being picked out by weathering and erosion along their edges. The rucksack gives an idea of scale.

- Evidence against this being tectonic (compressional) 'folding' of the bedding has in fact already been seen at localities 1 and 2. Can you think what it is?
- What non-tectonic origins for the sediment 'folds' could there be? [9]

Determine the nature of the Carboniferous–Permian boundary

The bedding only appears contorted within the ridge itself. The whole foreshore overlying it to the south is composed of quite different bed-forms. The two meet at a flat, planar unit on the southern edge of the ridge which marks the boundary between the Carboniferous and Permian Periods on this foreshore (see Figure (ii)). This contact is also locality 4 (Figure 2.7).

FIGURE 2.7 Looking down at the top bedding surface of the Carboniferous white ridge overlain by the basal Permian units. The position of this boundary is based on the apparently abrupt change in depositional environment. Closer inspection should yield important evidence on just how quickly this occurred. The length of the tape measure is about 30 cm.

- What are the dip and dip direction of the flat bed that marks the top of the Carboniferous? (At least two measurements should be collected and plotted on Figure 2.4) [10] [11]
- Compare the dip direction of the top Carboniferous surface in Figure 2.4 with that of the beds at locality 1. Is there any difference between the two?
- How does this affect the interpretation for the origin of 'folding' in the main ridge body?

- There is a possibility of dessication cracking on this top surface; can you see where this might be? [9]

Now examine the basal 20 cm of the Permian beds with a hand lens and compare the lithology with that of the ridge described earlier.

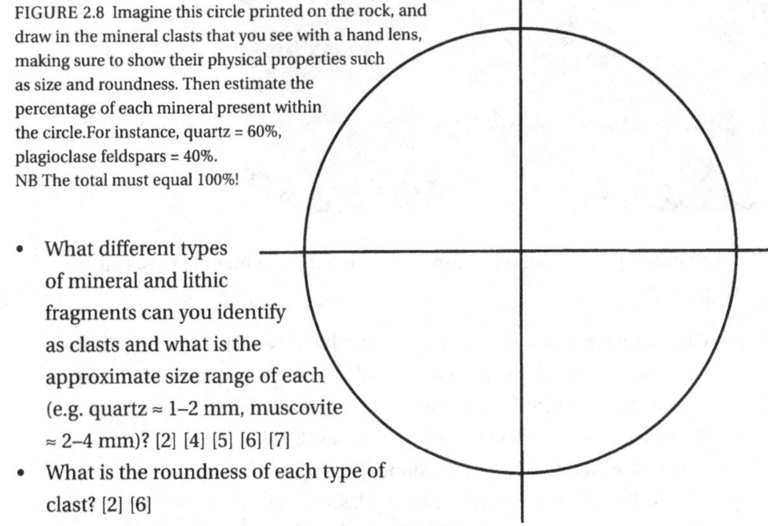

FIGURE 2.8 Imagine this circle printed on the rock, and draw in the mineral clasts that you see with a hand lens, making sure to show their physical properties such as size and roundness. Then estimate the percentage of each mineral present within the circle.For instance, quartz = 60%, plagioclase feldspars = 40%. NB The total must equal 100%!

- What different types of mineral and lithic fragments can you identify as clasts and what is the approximate size range of each (e.g. quartz ≈ 1–2 mm, muscovite ≈ 2–4 mm)? [2] [4] [5] [6] [7]
- What is the roundness of each type of clast? [2] [6]
- How well sorted is the overall sediment? [2] [6]
- What is the sphericity of each type of clast? [2] [6]
- Using Figure 2.8, what percentage of the 'whole rock' does each type of mineral represent? [2]
- The mixture of sediment here makes naming the rock a little difficult, but what is the most appropriate name for it? [2] [7]
- In what ways have the sediment composition and 'maturity' changed since locality 2? [2] [4] [6]
- Does the grain size, sorting and roundness give any clues as to whether these sediments were deposited from water or air? [2] [6] [9]

Reconstruct depositional environments of the Permian rocks

Having recorded many changing bed-forms and lithologies over a short stratigraphical distance from localities 1 to 4, the remaining Permian rocks can be dealt with as a single unit for study all the way to the harbour at [NS 0260 4280]. This succession is known as the 'New Red Sandstone' (see Figure (ii)). So walk a few metres into this unit to get an idea of the bed-forms

FIGURE 2.9
Contemplating bed-forms in the Permian units.
NB Beware of wet surfaces or those covered in the light green seaweed; both can be extremely slippery.

on an outcrop scale and then stop to examine the sediment in close-up (Figure 2.9).

- What different types of mineral can you identify as clasts and what is the approximate size range of each mineral 'species' (e.g. quartz ≈ 1–2 mm, muscovite ≈ 2–4 mm)? [2] [4] [6]
- What is the roundness of each type of clast? [2] [6]
- How well sorted is the overall sediment? [2] [6]
- What is the sphericity of each type of clast? [2] [6]
- Using Figure 2.10, what percentage of the 'whole rock' does each type of mineral represent? [2]
- What is the name of this rock? [2] [7]
- In what ways have the sediment composition and 'maturity' changed since locality 4?

FIGURE 2.10 Imagine this circle printed on the rock, and draw in the mineral clasts that you see with a hand lens, making sure to show their physical properties such as size and roundness. Then estimate the percentage of each mineral present within the circle. For instance, quartz = 60%, plagioclase feldspars = 40%. NB The total must equal 100%!

With some idea of how the sediment is changing up-section and perhaps with ideas about environment beginning to form, it is important to gather as much other evidence as possible before drawing some final conclusions.

- What is the most striking feature of the bedding in these Permian rocks (sketch some examples including an idea of their scale)? [2] [9]

These Permian rocks might show a particularly characteristic type of sedimentary bed-form (see [9]). But the only way to be sure is to take a series of structural measurements and see how the dip direction varies. As a general rule a 'statistically significant' data set must contain at least 30 results, so aim to gather at least this number of readings at approximately regular intervals along the foreshore towards the harbour. Plot these on Figure 2.4. *NB Be sure to measure true bedding and not just sloping weathered surfaces.*

- How can the spread of dip directions in the Permian units on Figure 2.4 be explained in terms of original depositional environment? [11] [9]
- Were these Permian sediments deposited in water or air? [9]
- How does this interpretation differ from that at locality 4?
- What 'current' direction do the Permian beds record as compared with that at locality 2 (measure and plot it on Figure 2.4 in the same way as for locality 2)?
- Is there any difference in the angle of dip between Carboniferous beds and the New Red Sandstone and if so, how might this be explained? [10] [9]

Synthesis

Five localities have been investigated during the course of this exercise, each revealing a significant variation in sedimentary bed-form and composition.

- What generalisation can be made about depositional environments at Corrie from the latest Carboniferous to earliest Permian (compare each locality with the next and think in terms of the depth in water)?
- From Figure 2.11, does the plotted position of Britain during the Permian fit the environmental evidence collected on Corrie foreshore? (Think of where similar environments or climates might be found on continents today at the same latitude).

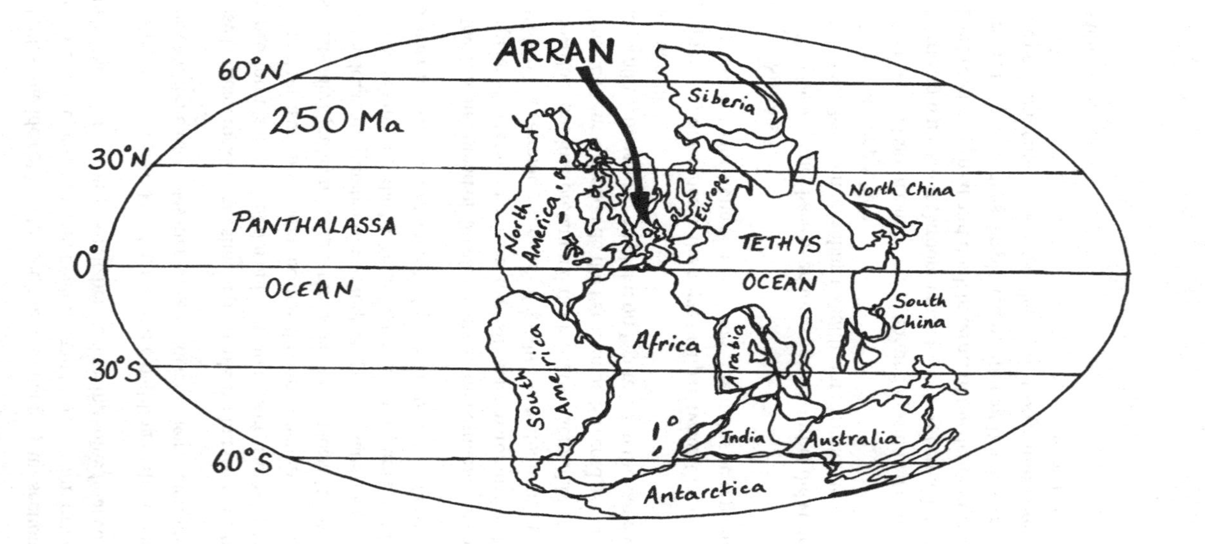

FIGURE 2.11 A simplified palaeogeographic map of the world during the Permian. Palaeomagnetic data suggest that Arran was at about 15° N during the deposition of the New Red Sandstone at the centre of a massive supercontinent called 'Pangea', and it then gradually moved north to its present position. This map is based on an outline in Long, J.A. (ed.) 1993, *Palaeozoic vertebrate biostratigraphy and biogeography*. Belhaven Press, London.

Exercise 2: field notes exact location	General location: Date:	data

Exercise 2: field notes exact location	General location: Date:	data

Exercise 2: field notes exact location	General location: Date:	data

Exercise 2: field notes exact location	General location: Date:	data

Exercise 2: field notes exact location	General location: Date:	data

Exercise 2: field notes exact location	General location: Date:	data

3 Intrusive igneous rocks *

(Tormore shore, west coast)

PRELIMINARIES

Task To assess evidence of composite igneous intrusions by returning to localities described in a set of rain-damaged field notes.

Logistics *Start Point*: A841 forestry car park south of Machrie [NR 8978 3130]

Tormore is a small hamlet on the west coast of Arran about 2 km south of Machrie and 4.5 km north of Blackwaterfoot. A large Forestry Commission car park at the roadside about 1 km south of Tormore provides access to the exercise area. From the car park, take the footpath signposted to 'King's Cave' which contours westwards around the northern flank of Torr Righ Beag and then turns south to follow along the cliff top. This 35 min walk gives a good view down onto the foreshore exercise area and many of the major composite dykes can be spotted before joining the beach at the ridge called An Cumhann (which marks the southern boundary of the area) [NR 8850 3134]. *The exercise is tide dependent*, especially in the crossing of An Cumhann. It is still possible from the north if caught out by the incoming tide, but only via an indistinct footpath directly up the cliff (at about [NR 8849 3132]) to join the main path on the top. Exit from the exercise area is possible at any tide along the foreshore to the north.

Length

4 hours: 35 min walk to begin first step, 2 hr 40 min to complete the exercise, 45 min walk back to *Start Point*

Field skills required [1] [2] [3] [4] [5] [10] **[12] [13] [14]**

Background information

The majority of igneous rocks found on Arran have intruded the surrounding country rock. This exercise introduces dykes, which are the most common type of intrusion on the island. Along the foreshore at Tormore these are particularly impressive as each one has formed from several periods of magma emplacement, often involving both acidic and basic compositions (Figures 3.1 and 3.2).

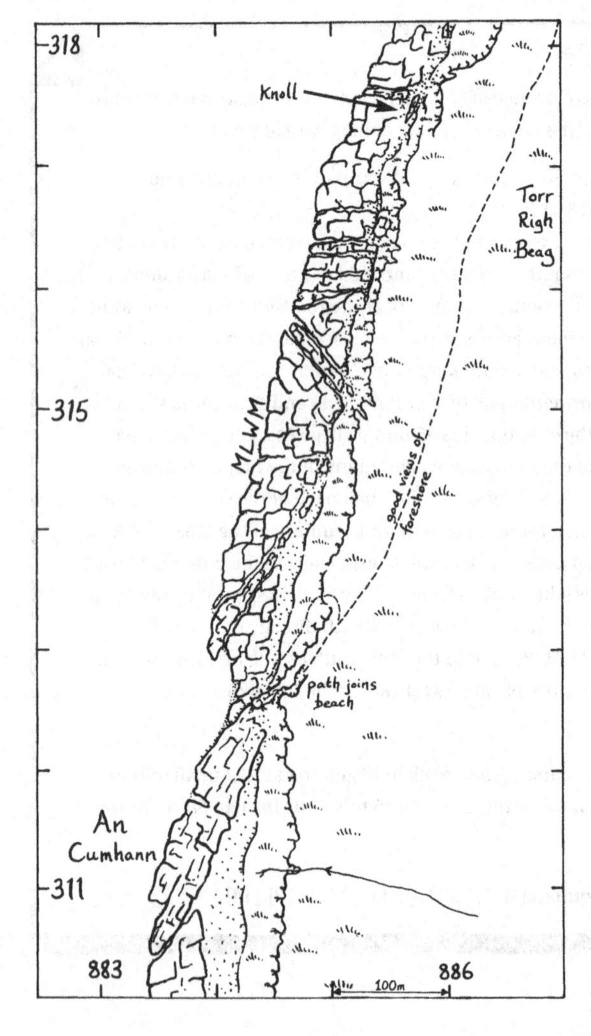

FIGURE 3.1
Exercise base map of Tormore foreshore. The southern and northern limits of the exercise area are marked as An Cumhann and knoll respectively. The Forestry Commission car park *Start Point* is about 1.5 km east of the map area.

FIGURE 3.2 Looking north across the composite dykes of the exercise area from the cliff above the An Cumhann ridge. The pitchstone knoll marking the northern limit of the area (shown on Figure 3.1) can be seen just projecting into the sea below the centre of the hill in the distance.

This exercise has two purposes. Firstly, it takes a closer look at these 'composite dykes' to find out more about their structure and how igneous intrusions might form. The field work requires observations on dyke composition and texture, including an introduction to igneous glasses. It requires assessing the influence of cooling rate on crystal growth and the viscosity of melts in producing flow banding or folding. Also, an important aspect of the work is how to determine the relative age of intrusions from field observation. Finally, the exercise concludes by considering how fractionation processes in the original magma chamber might have affected the resulting intrusions.

The second aim of this exercise is to teach the importance of field notes and sketches, which are used by a geologist to record primary field evidence. It is common for geologists to disagree with each others' interpretations, so it is an important requisite that all field evidence can be checked if necessary by another geologist (who might be unfamiliar with the area). This means that what you see in the field must be written down clearly and objectively, with as much information as possible about the location and nature of the evidence. The best way to illustrate this is to actually try and follow another person's field notes. The field exercise introduces Morag, a keen student of geology, and pages from her notebook of the Tormore shore. Unfortunately, heavy rain began to take its toll on Morag and her notes so they deteriorated during the day. Observing what vital information is missing should demonstrate the common pitfalls of note-taking and what might be done to avoid them.

exact location	General location: Tormore Shore Date: 21/7/96	data
Locality 1 [NR 8843 3135] Flat rock at edge of sea 100 m north of An Cumhann ridge	<u>*Observation*</u> • A black-green ridge exposed between low and high tides; strikes across the flat-lying rock of foreshore, bordered either side by a narrow gully • Ridge is not a simple dipping planar sheet, but has a shape in cross-section • Ridge has curving planar structures within it and apparently dipping upper and lower surfaces • Conchoidal fracture to ridge rock • No crystals or grains seen in central region, but is lighter in colour and crystalline at edges • Will not scratch with metal blade • Both gullies contain the same poorly exposed, brown-green rock • It will not scratch with metal blade and appears to be microcrystalline <u>*Interpretation*</u> • Ridge composed of acidic igneous glass; it is a pitchstone • Lighter, crystalline edges to ridge are devitrifying pitchstone; producing a felsite • Ridge is discordant with surrounding country rock; it is intrusive and a dyke • Curving planar structures are flow folds produced during injection • Shape in profile produced by resistance of horizontal country rock to intrusion from below • Gullies are parallel basaltic-andesitic dykes flanking the central pitchstone	trend of ridge: 020° ridge 4.7 m wide *Figure 3.4a* *Figure 3.5a* gullies ~20 cm wide dip of ridge-gully-contacts: 60/140

FIGURE 3.3 Morag's first field note page describing the geological structure of the dyke at locality 1.

THE FIELD EXERCISE

Read through all the instructions thoroughly before beginning work. Annotate maps and figures as required and write your answers to the questions as field notes on the pages provided at the end of the exericse.

Steps needed to achieve the set task

1 Examine the overall structure of composite dykes.
2 Investigate composite dyke mineralogy.
3 Determine relative ages in parallel dykes.
4 Determine relative ages from cross-cutting relationships.
5 Synthesis.

Examine the overall structure of composite dykes

To start the exercise, take the footpath to 'King's Cave' from the Forestry Commission car park (*Start Point*) and after about 35 minutes this drops down onto the beach via a gully through the large composite dyke of An Cumhann [NR 8840 3115]. This dyke forms a ridge that runs south-west out to sea. Walk onto the ridge until there is a clear view northwards along the shore; this is the exercise area (refer again to Figure 3.2). The field notes from Morag's first locality are shown in Figure 3.3 **[12]**. Read through them and use the information she gives to find the outcrop [1] and then try to identify her field evidence recorded as *observation*.

The notes of Figure 3.3 refer to two photographs of some particularly interesting features to be seen in this composite dyke (Figures 3.4a and 3.5a). However, a geologist seldom has the luxury of being able to annotate a photograph with geological observations whilst still in the field. This can be done later when the film is developed, but by then of course the observations will have been forgotten. The answer to this problem is to make a field sketch. So use Figures 3.4b and 3.5b to draw field sketches of the view shown in each of their corresponding photographs (Figures 3.4a and 3.4b respectively) **[13]**. Use Morag's notes and your annotated field sketches to answer these questions.

- Is this a dyke or a sill, and why?
- What would a liquid be like that could produce folds in itself as it flowed? **[14]**
- An acidic melt (with a high silica, SiO_2, content) is particularly viscous; how does this compare with the compositional identification of the black–green ridge? [5]
- What does the presence of an igneous glass suggest about cooling speed? [5]
- How does this answer compare with that for the origin of flow folds; could the rock of the central ridge have been chilling as they formed? [5] [14]

FIGURE 3.4(a) Looking back south towards An Cumhann from locality 1. The flat-lying Triassic sandstones of the foreshore contrast with the truncating pitchstone intrusion. The hammer gives an idea of scale.

Location: 10 m WSW of locality 1 *Grid Ref:* NR 8842 3134 *Facing:* 190°

Scale:

(b) Blank field sketch template for the view seen in (a).

FIGURE 3.5(a) Superb flow-folding in the pitchstone ridge near locality 1.

Location: 30 m SW of locality 1 *Grid Ref:* NR 8840 3132 *Facing:* ~120°

Scale:

(b) Blank field sketch template for the view seen in (a).

exact location	General location: Tormore Shore Date: 21/7/96	data
Locality 2 [NR 8850 3154] Flat-topped ridge striking across foreshore ~200 m NNE of locality 1	*Observation* • Two wide, vertical-sided, parallel gullies flank large, flat-topped ridge • Ridge strikes over foreshore towards cliff where it forms a tall, vertical slab • Country-rock hard and whitened at northernmost margin • Low-lying, narrow, linear exposure of black-green to light-grey rock of variable thickness separated from country-rock to north by gully • Where rock is black-green it has a conchoidal fracture	trend of ridge: 140° *Figure 3.7* gully varies in width from 40 cm to 1 m
towards sea	• Where light-grey, rock has phenocrysts in microcrystalline groundmass • There is parallel banding at either margin, particularly the southern • Some reddening around this banding • Southern margin in intermittent contact with low, eroded outcrops of a green-grey, fine to medium crystalline rock occupying another gully • Rock in gully more jointed towards main ridge at southern contact • Sharp jump of 0.5 m out of gully onto main ridge to south • Main ridge composed of green, finely crystalline rock with phenocrysts up to 1 mm in size • Oblique-running joints across ridge • Drop down into gully on southern margin of ridge • This gully contains sporadic outcrop and has variable width	width of ridge 1 m ± 20 cm gully 2.9 m wide main ridge 6.8 m wide gully ~1 m ± 2 cm wide
~20m down from edge of grass	• Outcrop stretches across gully to show homogeneous composition; brown-green microcrystalline, with prominent white veining • This is in contact with white, hardened sandstone to the south *Interpretation*	

FIGURE 3.6 Morag's second field note page describing the dyke at locality 2.

Investigate composite dyke mineralogy

FIGURE 3.7 Locality 2 looking towards the cliff. Note the vertical slab of rock in the cliff-top which demonstrates the trend of the dyke and its dip as a vertical, planar sheet. Also, the central ridge can be seen to be flanked by two vertical-sided gullies. Again, the hammer gives an idea of scale.

Morag's second locality was the next composite dyke to the north, which displays greater variations in lithology. So walk the 200 m north–north-east to her locality 2, described in Figures 3.6 and 3.7. Unfortunately, her page of interpretations is missing, presumably destroyed by the rain. Nevertheless, read through the observations that survive and try to locate them on the outcrop. These appear to show that Morag made a transect over the dyke from north to south, perpendicular to its trend. What she records is a rather long and not particularly 'user friendly' list of observations. A sketch of the dyke in cross-section would have summarised them much more clearly and succinctly. Rather conveniently, Figure 3.8 shows just such a profile so annotate it with the observations and interpretations you make to answer the following questions concerning Morag's notes.

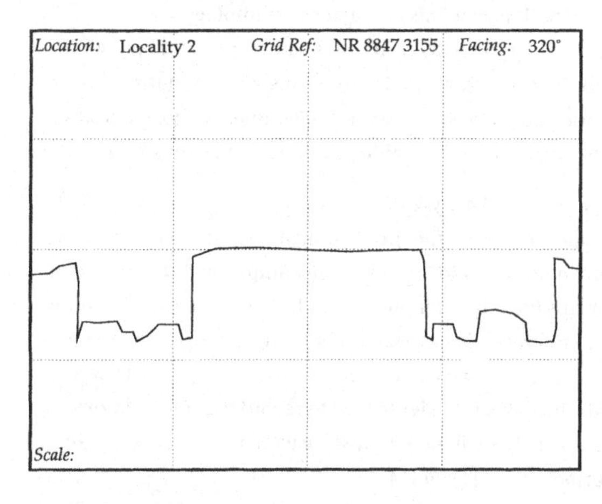

FIGURE 3.8 Blank sketch section through the composite dyke at locality 2.

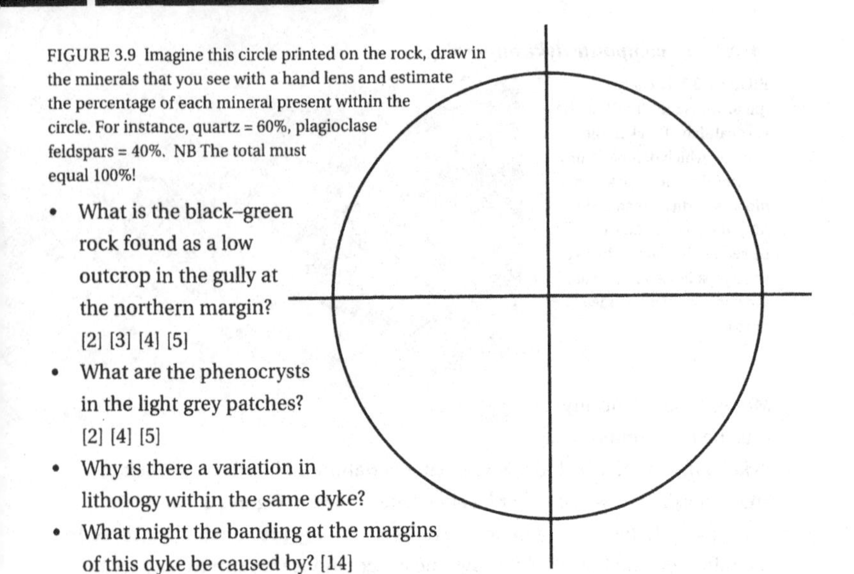

FIGURE 3.9 Imagine this circle printed on the rock, draw in the minerals that you see with a hand lens and estimate the percentage of each mineral present within the circle. For instance, quartz = 60%, plagioclase feldspars = 40%. NB The total must equal 100%!

- What is the black–green rock found as a low outcrop in the gully at the northern margin? [2] [3] [4] [5]
- What are the phenocrysts in the light grey patches? [2] [4] [5]
- Why is there a variation in lithology within the same dyke?
- What might the banding at the margins of this dyke be caused by? [14]
- The reddening around these features is odd; what might it be due to? [14]
- What is the green–grey, fine–medium crystalline rock occupying the gully on the north margin of the main ridge? [2] [3] [4] [5]
- What is the groundmass composition of the main ridge? [2] [3] [4] [5]
- Can the phenocrysts be identified, and if so, what mineral are they? [2] [4] [5]
- Do these phenocrysts have a similar SiO_2 composition as the groundmass? [2] [5] (Use Figure 3.9 to illustrate this characteristic lithology)
- What is the rock occupying the gully south of the main ridge? [2] [3] [4] [5]
- Look at your annotations to Figure 3.8. Is there a discernable pattern to the position of acidic or basic intrusions within this 'composite dyke', and if so, what is it (e.g. basic intrusions in the middle, acidic at the edges)? [5]

Determine relative ages in parallel dykes

Clearly the rain had set in heavily by the time Morag made it to her third locality to examine the relative timing of intrusion in the composite dykes. Much of the field note page for this locality is missing, including some of the directions to the outcrop (the grid reference is only six rather than eight figures; Figure 3.10). Again, the best way to summarise your observations here is to draw a sketch section through the dyke. Use Figure 3.12 to record the information missing from Morag's notes that will help to resolve which of the two interpretations she gives is correct [2] [3] [4] [5] [14].

exact location	General location: Tormore Shore Date: 21/7/96	data
Locality 3 [NR 885 316] Ridge on south side of sandstone knoll 10 m W of cliffline	*Observation* • Flat ridge bounded by parallel gullies on either side • Parallel banding on northern margin of central ridge	trend of ridge: *Figure 3.11*
	Interpretation Two opposing possibilities for the relative ages of emplacement in this composite dyke: 1. A single, basic dyke intruded first, followed by the felsite which has come up in the centre of it to push it apart; flow-banding in felsite demonstrating emplacement against already (partially) cooled basic intrusion either side 2. Felsite intruded first, flow-banding against sandstone margins, then basic dykes intrude up edges to push sandstone and felsite apart	

FIGURE 3.10 Morag's third field note page describing the dyke at locality 3.

FIGURE 3.11 Locality 3 looking east at the cliff. The sandstone knoll mentioned in Morag's notes can be seen just left of centre at the top of the beach.

- Are the dykes on either side of the ridge the same composition and how might this help resolve the timing of dyke emplacement? [2] [4] [5] [14]
- Are there any chilled margins to help decide which intrusion has cooled against which? [14]
- Does your assessment of the field evidence support Morag's 'interpretation 1', or her 'interpretation 2' (Figure 3.10)?

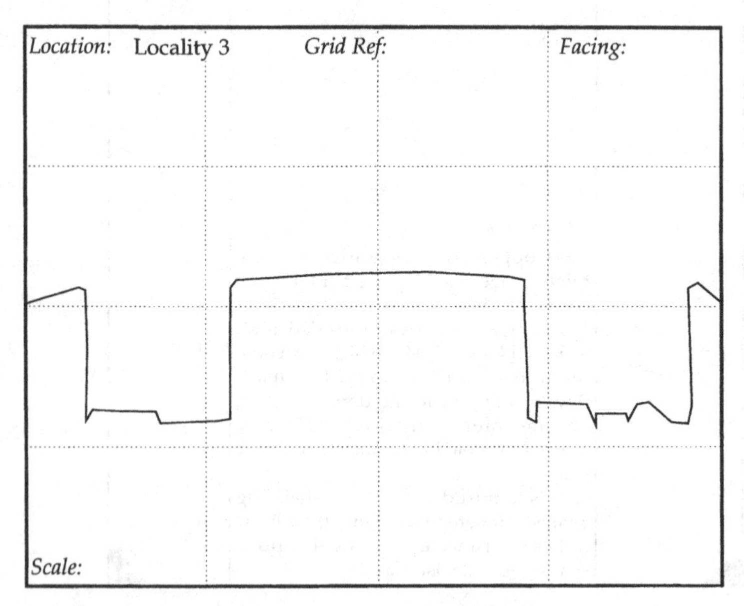

FIGURE 3.12 Blank sketch section through the composite dyke at locality 3.

exact location	General location: Tormore Shore Date: 21/7/96	data
Locality 4	<u>*Observation*</u> • My feet are wet and it's still raining • Looked at cross-cutting relationship between two dykes at a large, upstanding pitchstone knoll on foreshore	*Figure 3.14*

FIGURE 3.13 **Morag's final field note page describing very little at locality 4.**

Determine relative ages from cross-cutting relationships

It seems that Morag gave up shortly after reaching locality 4. Here, instead of her notes being partially destroyed they seem never to have been written in the first place (Figure 3.13). The only information recorded is a jumbled sentence of location, observation and interpretation. Luckily, there is a 'knoll' on the foreshore north of Morag's locality 3 which is easy to spot from a distance and would seem to fit well with what little description she gives (its position is marked on Figure 3.1). So walk the ~150 m north–north-east from locality 3 to this knoll, crossing a couple of thin basic dykes on the way. Make a note of the grid reference and nearest landmarks before examining the composition of the knoll and the rest of the intrusion as it heads toward the sea. Make the necessary observations to confirm, hopefully, that it is a pitchstone.

- What is the trend of the dyke which includes the pitchstone knoll? [10] [14]
- Is this dyke in direct contact with the sandstone on either side?
- What is the dip of the dyke contacts? [10] [14]
- What structures are in the pitchstone knoll? [14]

After this brief description of the first intrusion, hunt around the near vicinity to try and find evidence of the other dyke that Morag thought she saw trending to cut across it (Figure 3.14).

- What is the composition of this second dyke? [2] [3] [4] [5]
- What is its trend and the dip of its contacts? [10] [14]
- Are there any structures in it similar to those in the 'pitchstone knoll dyke'? [14]

FIGURE 3.14 The 'pitchstone knoll' of Morag's locality 4 seen from the edge of the grass and looking NNW. In the foreground is a mysterious vertical ridge of hardened sandstone trending towards the sea.

Look for outcrops in the area where the two dykes should cross and whether the course of one has been off-set during the emplacement of the second [14]. Frustratingly, outcrop at the critical junction of the two is all but missing. However, the relationship of the two can be resolved in another way. Return to the seaward end of the 'knoll' dyke and walk along it towards the cliff trying to step only on pitchstone.

- Is the trend of the 'knoll' dyke the same throughout all of its course across the foreshore and into the cliff? [10] [14]
- Where precisely does it change trend? [10] [14]
- How does the composition of the two dykes compare? [2] [3] [4] [5]
- What structures in the two dykes suggest that they may in fact be off-shoots of the same one? Draw the interpreted course of these 'two' dykes onto Figure 3.1.

Synthesis

Although different aspects of dyke structure and composition have been studied at four localities, it is perhaps their origin which is most intriguing. To find intrusions of acidic and basic composition in such complex relationships is a relatively rare phenomenon. Consequently, the Tormore dykes have been a source of curiosity to geologists since their first (and remarkably accurate) description by Robert Jameson as long ago as 1800. The most famous study of them was by John Wesley Judd in 1893 and they are often known colloquially as 'Judd's Dykes'.

To intrude dykes of different composition along the same fracture does not necessarily need separate magma chambers. It is possible to *fractionate* molten magma into different compositions within the same chamber. This can be achieved in many different ways, but one of the simplest is to grow crystals in it. Any body of magma moving upwards in the crust will cool as it does so. Each mineral 'species' has a different melting point; for ferromagnesian minerals (in basic magma) it is very high, whilst SiO_2-rich minerals (in acidic magma) melt at a much lower temperature. Of course, this principle works the other way round too and this is where it becomes applicable to an igneous intrusion; ferromagnesian minerals crystallise out first in a cooling magma chamber, but the SiO_2-rich minerals can only solidify when the temperature gets much lower (see [5]). Taking chemical elements out of the cooling liquid magma in order to build solid crystals will obviously change the composition of the melt left behind. Another way of having two separate compositions together in the same chamber might be if they are both still liquid, but *immiscible*. Imagine a mixture of oil and water in the same beaker. The oil will float on the water because it is less dense and will not mix with it. Both acidic and basic magmas have large density and viscosity differences and so might be expected to behave immiscibly if contained together in the same magma chamber.

Consider these ideas and imagine a large magma chamber sitting kilometres below the Tormore shore. The cracks or fissures produced above it as it slowly rises have allowed some of the melt to escape up towards the surface to give the succession of parallel dykes now exposed by erosion. Look over the field evidence contained in both Morag's notes and yours once again.

- How might fractionation of melt in the magma chamber help to explain the differences and patterns of composition across the composite dykes? [5] [14]
- How might this fractionation affect the order in which melts of acidic and basic compositions are intruded? [5] [14]

Exercise 3: field notes exact location	General location: Date:	data

Exercise 3: field notes exact location	General location: Date:	data

Exercise 3: field notes exact location	General location: Date:	data

Exercise 3: field notes exact location	General location: Date:	data

4 Changing the record: metamorphism and deformation **

(Rubha Airigh Bheirg, north-west coast)

PRELIMINARIES

Task To recognise phases of deformation in the Dalradian metasediments and attempt to work out their sequence of formation.

Logistics *Start Point*: The promontary of Rubha Airigh Bheirg [NR 8855 4785]

Rubha Airigh Bheirg (pronounced badly in English along the lines of 'Rue-very Veg'; with a hard 'g') is an isolated, flat promontary approximately 2.5 km south-west of Catacol on the north-west coast of Arran. Vehicles can usually pull off the road onto the flat grass of the headland for the duration of the field work [NR 8855 4785]. From here there is immediate access to the outcrops of the exercise area, which stretch south-west along the foreshore for about 250 m. If travelling by bus and the driver appears in cheery mood then ask to be let off at the exercise area itself (explain that it is near the old grave yard at Lennimore or North Thundergay). Failing which, Catacol is the nearest official bus stop and the 2.5 km to the headland will have to be walked along the roadside. Unfortunately, the walk back will have to be made at the end of the exercise as well unless a passing bus is successfully stopped (perhaps by pretending to be a sheep in the road). *The exercise can be attempted whatever the state of the tide, although some small-scale structures and folds are covered when it is high.*

Length

2 hr 30 min: 2 min walk to begin first step, 2 hr 20 min to complete the exercise, 8 min walk back to *Start Point*

Field skills required [1] [2] [3] [4] [6] [7] [9] [12] [13] **[15] [16] [17] [18] [19] [20]**

Background information

After burial, Neoproterozoic to Cambrian age deep-water sediments were subjected to several cycles of heating, compression and uplift before finally being exposed as the Dalradian Supergroup across the northern half of Arran (see Figures (i) and (ii)). In essence, these processes demonstrate the second half of the rock cycle investigated in Exercise 1. Particularly good examples of the tectonically deformed Dalradian 'metasediments' can be found at Rubha Airigh Bheirg, with many folds eroded to display their three-dimensional structure.

This exercise begins by introducing some of the features of Arran's metamorphic rocks, including textures, growth of metamorphic minerals and different varieties of cleavage. It then sets out to investigate how a geologist can unravel the deformation history of rocks such as these by the recognition of successive deformation 'phases', with each new one 'overprinting' the results of the last. This is best illustrated at Rubha Airigh Bheirg by the folding developed in the metasediments. Consequently, much of the exercise involves collecting fold plunge measurements with which to try and resolve how many phases of compression have taken place. At first glance the data appear random, but by plotting on a stereonet the pattern should soon become clear. The exercise concludes by combining these fold data with earlier observations on cleavage and the original bedding to draw up a table summarising the deformation history of the Dalradian in this part of Arran.

THE FIELD EXERCISE

Read through all the instructions thoroughly before beginning work. Annotate maps and figures as required and write your answers to the questions as field notes on the pages provided at the end of the exericse.

Steps needed to achieve the set task

1 Identify characteristic features of Dalradian deformation.
2 Measure fold plunges at north-east road hump.
3 Measure fold plunges at central road hump.
4 Measure fold plunges at south-west road hump.
5 Synthesis: deformation phases.

Identify characteristic features of Dalradian deformation

The best exposures on which to begin investigating metamorphic rocks and their textures are those closest to the *Start Point* on the foreshore (Figure 4.1). So begin by walking south-west from Rubha Airigh Bheirg for ~100 m, joining

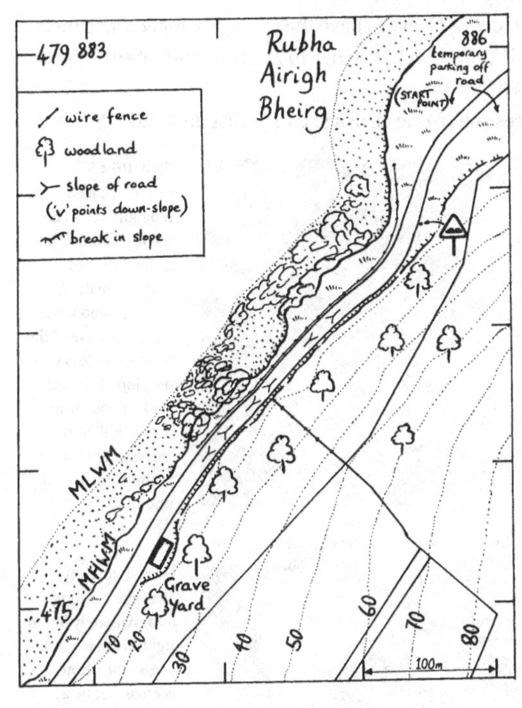

FIGURE 4.1 Exercise base map of Rubha Airigh Bheirg showing the three road humps which are the best way of locating yourself on the outcrops. The exposure disappears towards the grave yard to the south-west of the exercise area.

the beach where outcrop first appears [NR 8850 4777] [1]. *Be careful* when crossing these rocks as they can be slippery in places. Examine the foreshore metasediments from here to opposite the first hump in the road about another 70 m to the south-west (this area is shown in Figure 4.2, from [NR 8850 4777] to [NR 8843 4772]). In many of these outcrops it is difficult to find evidence of the original bedding because this has been 'overprinted' by the development of metamorphic cleavage planes which are now the dominant planar fabric in the rocks **[15]**.

FIGURE 4.2 Looking north at the area of investigation for the first step, with the *Start Point* in the background at Rubha Airigh Bheirg.

- How many different types of cleavage have the rocks developed since their deposition and burial? Draw quick sketches to illustrate your observations (see Figures 4.3 and 4.4). [15]
- Are there any instances where one cleavage cuts across another? [15]

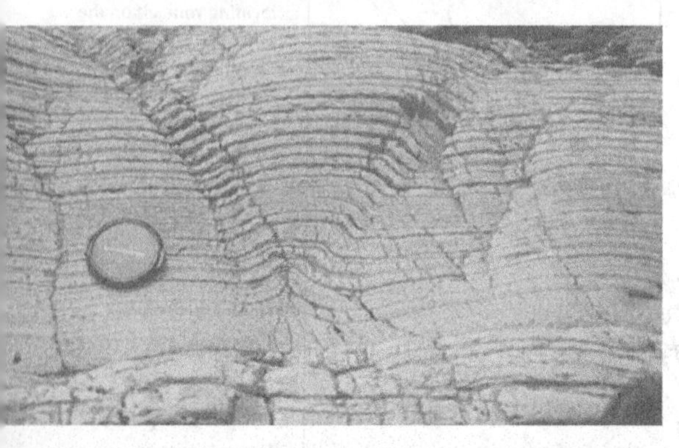

FIGURE 4.3 Closely-spaced repetative cleavage planes (running left to right) in finer-grained 'beds'. A coarser band is visible at the base of the photo with less well-developed cleavage. Kink bands forming a 'box fold' in the cleavage are present in the centre of the photo.

FIGURE 4.4 Deformed 'slaty' cleavage running approximately left to right in cross-section across the photo is cross-cut by a near-vertical axial planar cleavage. The coin gives an idea of scale.

These metamorphic rocks are thought to have been deep-water clastic sediments originally.

- Can you find any evidence of grains or clasts in them?
- Are there any bands running through the rocks that contain predominantly coarse or fine clasts which might indicate former sedimentary bedding (see Figure 4.3)? [6] [9]
- 'Dalradian' is the stratigraphic name given to this 'Group' of metamorphosed sediments, but what is the proper lithological name for the rock at this locality (use Figure 4.5 to sketch examples of the texture)? [2] [4] **[16]**

- What might have been the name of the original sedimentary rock? [7] [16]

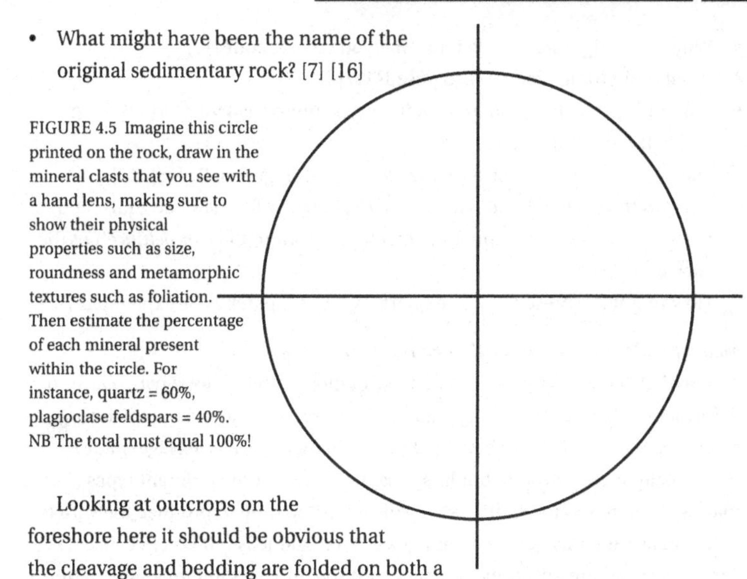

FIGURE 4.5 Imagine this circle printed on the rock, draw in the mineral clasts that you see with a hand lens, making sure to show their physical properties such as size, roundness and metamorphic textures such as foliation. Then estimate the percentage of each mineral present within the circle. For instance, quartz = 60%, plagioclase feldspars = 40%. NB The total must equal 100%!

Looking at outcrops on the foreshore here it should be obvious that the cleavage and bedding are folded on both a small and a large scale. It may appear that these are fairly randomly orientated folds, but in fact they show a distinct pattern. Before setting out to see what that might be, it is worthwhile just spending a few minutes in describing the fold types and their characteristic geometry **[17]**. So look around the outcrops in this short stretch of foreshore and try to find as many different examples of folding as you can (Figure 4.6). Make quick sketches of each to show their general shape and structure [13] [17].

FIGURE 4.6 A variety of fold geometries on the foreshore at the first locality. Folding on a ~1 m scale can be seen throughout, but also notice that the rocks on the left appear to dip out to sea, whereas those on the right dip towards the road. A closer look at the outcrops should also reveal many folds on a centimetre scale.

- What is the tightness of the folds that you have found? [17]
- What fold profiles can be recognised? [17]
- Can a vergence direction be seen in any asymmetrical folds and if so, in which direction? [1] [17]
- How does the folding relate to the cleavage(s) (Figure 4.4)? [15] [17]
- How do the different fold types relate to coarse or fine-grained units, and what does this suggest about the mechanical strength (*'competance'*) of the rock in each case?
- How might the presence of the quartz veins and patches be explained? [17]

Measure fold plunges at north-east road hump

The field observations so far should have demonstrated *at least* two 'phases of deformation' because a cleavage must have formed in the sediments during an earlier event in order to be folded in another later. This cleavage–folding relationship is easy to spot, but less so is that between the different types of fold; were there separate, successive episodes of folding to produce the variety seen? Answering this question with observation alone is difficult because it is very easy to become disorientated amongst the outcrops and miss any patterns that may be present to their axis directions (see [17]). However, all is not lost. By measuring the *plunge* of folds ([17] **[18]**) and seeing how they relate to each other when plotted on a spherical graph, called a stereonet, this problem can be solved ([**19**]). A stereonet can show which folds were formed in the same deformation episodes, how many episodes there were, and more important, their compression directions can also be calculated from it. Consequently, the remainder of this exercise is concerned with collecting and plotting enough fold plunge data to resolve the fold orientations before a final synthesis of all the field evidence.

FIGURE 4.7 Looking north towards Rubha Airigh Bheirg at the road humps that can be used as landmarks for the three sectors of fold plunge data collection opposite them on the foreshore.

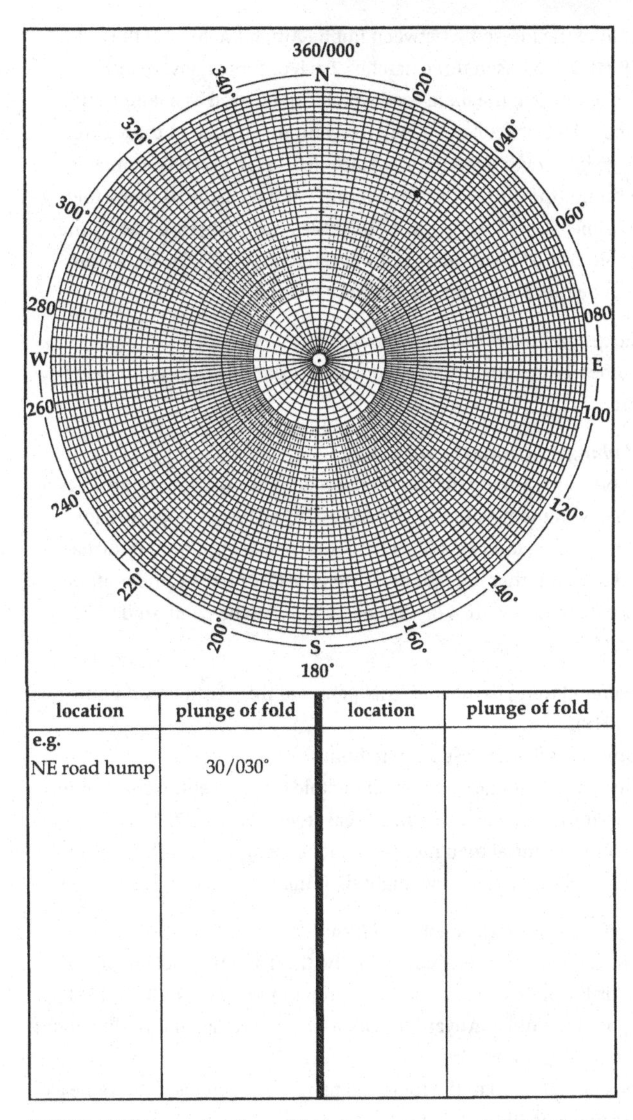

location	plunge of fold	location	plunge of fold
e.g. NE road hump	30/030°		

FIGURE 4.8 Write your location and fold plunge measurements in the columns provided. Then, using [19] as a guide, put a dot for each datum point on the stereonet. An example of how to plot a single value of a fold plunging 30/030° is shown to help you. Try and plot data from each road hump in a different pencil colour as this will highlight any changes in fold orientations between them.

Exposure along the foreshore between Rubha Airigh Bheirg and the old graveyard [NR 8835 4754] is in three patches. Each of these conveniently corresponds to a bump in the main road which can be used as a landmark (Figures 4.1 and 4.7). Data collection will concentrate on each of these three areas. So now go back over the outcrops examined so far which lie opposite the first road hump (the 'north-east road hump') and measure as many fold plunges as possible; about ten should be possible here ([17] [18]). Use Figure 4.8 to record and then plot your data onto the stereonet [19]. The data from this locality can be quickly summarised before moving on.

- What is the general trend of the fold axial planes? [17] [18]
- In which general direction are the folds plunging and by approximately how many degrees? [17] [18]

Measure fold plunges at central road hump

Walk over the break in outcrop just below the road for only about another 40 m to come parallel with the next hump [NR 8840 4767]. The outcrops lie a lot lower on the beach here but can still yield plenty of vital structural information. Collect at least another ten fold measurements using the same method as before and plot them on Figure 4.8. These readings can be compared with those at the last locality.

- Is there a general trend to the fold axial planes at this central road hump sector? [17] [18]
- Is it the same as at the north-east road hump?
- What is the general plunge direction of the folds at this central road hump sector and is it the same as at the north-east road hump? [17] [18]
- If the data for the central road hump do not cluster exactly with those of the north-east road hump, then how might their difference be explained? [19]

Measure fold plunges at south-west road hump

To finish the data collection, continue on to the third and final sector to the south-west. This is opposite another small hump in the road [NR 8837 4764] and within sight of the old graveyard further to the south. Again measure about another ten fold plunges and plot them on Figure 4.8.

- What is the general trend of the fold axial planes here, and is it the same as at the other two localities? [17] [18]
- What is the general plunge direction of the folds at this south-west road hump sector and how does it relate to that of the other two localities? [17] [18]

- Folds formed during the same deformation phase should all plunge in the same direction. So how can the difference in plunges shown in Figure 4.8 be best explained (think about how compression from different directions in separate deformation episodes might have given rise to the pattern in the data)? [19]

Synthesis: deformation phases

Being able to resolve how many phases of deformation rocks have undergone and in what order they might have occurred, is an important geological skill. So as a synthesis, an attempt can be made to do this for the Dalradian metasediments along the foreshore at Rubha Airigh Bheirg. The key to unravelling the metamorphic history of rocks such as these is to think of it as a linear jigsaw puzzle that is only one piece wide. At one end of the line you start with the original horizontal sedimentary bedding, and at the other you end with the rocks as you see them now. There is only one place in the line that each 'piece of deformation' will fit into and make sense, so it is a case of getting them in the right order. Field skill **[20]** demonstrates how you can do this in practice.

Each piece of the puzzle is given a label with a number to help place them all in the right order. For instance, any deformation that results in the formation of some kind of planar fabric, such as cleavage, is given the letter **'S'** (for 'planar **S**tructure') and then a number to show when it occurred (for instance S1 for a cleavage that happened first, S2 for the second one, S3 for the third and so on). Each different period of folding is given a similar label, but this time using the letter **'F'** (for '**F**old phase') and then a number for when it happened (F1 for folds that were the first to form, then F2 if these have been folded themselves and so on). Finally, the pieces labelled 'F' and 'S' can be listed together in order of when they happened and if they happened during the same period of deformation. Each deformation phase is again given a letter ('**D**' for '**D**eformation phase') and a number for whether it came first, second, third or later.

So consider all the field evidence gathered in this exercise, and make a list of all the different types of cleavage you have seen [15]. Then, make a list of the different fold phases observed or shown by Figure 4.8 [17] [19] [20]. As a general rule of thumb; folds with parallel axes and the same plunge will have formed at the same time, but if they have parallel axes and opposing plunges then they have been refolded. Finally, use Figure 4.9 to put each piece of field evidence in what you think is the right order to reconstruct the history of Dalradian metasediment deformation.

Planar foliation (S_1, S_2)	Fold phase (F_1, F_2)	Deformation phase (D_1, D_2)	Description of events that took place
S_0	-	-	Deposition of sedimentary beds
		D_1	

FIGURE 4.9 Chart summarising the deformation history of the Dalradian metasediments at Rubha Airigh Bheirg. The chart starts with the first event to have taken place, the deposition of the original sediment (this original planar fabric is given the label 'S', but has no number as it was not formed by deformation). From this point onwards you must write in each column what you think happened and in which order (see [20]). By the time you reach the bottom of the chart you should have recorded all the different types of cleavage and folding that you have seen during the exercise.

Exercise 4: field notes exact location	General location: Date:	data

Exercise 4: field notes exact location	General location: Date:	data

Exercise 4: field notes exact location	General location: Date:	data

Exercise 4: field notes exact location	General location: Date:	data

5 3-D thinking in time and space * *

(Glen Sannox, north-east coast)

PRELIMINARIES

Task The unscrupulous mining company *McNicholas Conglomerated* has bought up the derelict barytes mine in Glen Sannox. On contract to the company, your task is to suggest a new strategy to re-open the mine and achieve maximum production at the cheapest costs.

Logistics *Start Point:* Sannox Bay car park and telephone box [NS 0161 4540] Glen Sannox is on the north-east coast of Arran and the nearest village to the exercise area is South Sannox [NS 018 452]. The *Start Point* is the same as for Exercise 1, so its position and also that of the barytes mine can be found in Figure 1.2. Access to the exercise area is via the signposted cart track into Glen Sannox which can be found on the opposite side of the road from the *Start Point*. The mapping area approximates to grid square NS 00 45. *It may be useful to carry a calculator for this exercise.*

NB All vertical shafts and the larger collapsed underground 'quarries' (called 'stopes') in Glen Sannox are fenced or sealed off and on no account should you attempt to gain entry to them. The mine entrances on the hillside (called 'adits') and some smaller stopes remain open but again you should not enter the old workings. Many contain cracked and rotting pit props which demonstrate that all are unsafe and the majority are flooded. All shafts, adits and stopes found during a survey of the area for this exercise are marked on the map and it is best to take care when approaching them, particularly whilst descending a slope.

Length

3 hr 30 min: 15 min walk to begin first step, 3 hr 20 min to complete the exercise, 15 min walk back to *Start Point*

Field skills required [1] [2] [3] [4] [6] [7] [10] [12] [13] [14] [17] **[21] [22] [23] [24] [25]**

Background information

This exercise concentrates on applied geology and the potential clash between beginning a major mining operation and the ensuing environmental damage. In order to suggest a new mining strategy for Glen Sannox, field observations must be used to decide what the barytes veins look like in three-dimensions. Reconstructing the 3-D geological structure of an area is a common geological problem and is conventionally solved by combining a plan view (i.e. a map) with a vertical slice through the rock (i.e. a cross-section). So this exercise introduces simple geological mapping and constructing cross-sections of hydrothermal veins to predict the three-dimensional structure below ground. Once this has been done it investigates the mining techniques that could be used to extract the mineral veins based on calculations of the ore yield. Finally, the exercise considers the economics of re-opening the mine in conjunction with the environmental impact, not just on the mining area but also to the surrounding countryside and inhabitants.

Barytes ($BaSO_4$) is a bright white mineral belonging to the orthorhombic crystal system, with a tabular habit and three good cleavages (see [4]). It is used today as a paper filler, in barium-based chemicals and as a white pigment in paint. Barytes has a high density of 4.5 g/cm^3 and because of this it is also employed as a dense drilling mud to keep the pressure up in oil wells. In Glen Sannox it occurs as veins running through sands and conglomerates of the Old Red Sandstone. This occurrence suggests that the barytes is a precipitate from hydrothermal fluids. Barytes is generally insoluble in water, but this can be counter-acted by the presence of chlorides and CO_2.

Glen Sannox was first worked for barytes on a small scale from about 1836 to 1862, reaching a maximum annual output in the year production stopped of 742 tonnes. A second phase of development took place from 1918 until 1934, during which time the output soared to 8833 tonnes by the final year. The miners worked with pick and shovel along the narrow veins in what must have been very cold, wet and cramped conditions. A self-acting railway delivered wagons of unsorted rock from the adit on the northern side of the stream to the waterwheel-powered processing plant by Sannox Burn. Foundations of the mine buildings can still be seen. The rock was crushed and hand sorted on site before being taken by a second self-acting railway to storage bins on the coast at Sannox. Timbers from the loading pier can still be seen at low water; the barytes was destined for a mill at Glasgow. The mine workings themselves consist of a series of horizontal or gently inclined levels along which wagons could be pulled, connected by shafts (which were nearly vertical and mainly used for ventilation) and occasionally chutes through which the ore could be poured into wagons below.

THE FIELD EXERCISE

Read through all the instructions thoroughly before beginning work. Annotate maps and figures as required and write your answers to the questions as field notes on the pages provided at the end of the exericse.

Steps needed to achieve the set task

1 Describe barytes in outcrop.
2 Find the dip of the veins to the country rock.
3 Construct a map of the veins at the surface.
4 Project the veins below ground on a cross-section.
5 Synthesis: mining strategy.

Describe barytes in outcrop

Before the barytes veins can be mapped across the exercise area it is important to become familiar with the mineral's physical properties in hand specimen. A good locality at which to do this is at a vein exposed in the bed of Sannox Burn, locality 1 [NS 0064 4526] (Figure 5.1). The map for Exercise 1 (Figure 1.2) shows the *Start Point* at the Sannox Bay car park and also some hatched shading which corresponds to the approximate area covered by Figure 5.1, so use this map to navigate into the glen. Follow the signposted track into Glen Sannox opposite the *Start Point* for about 1 km (about 20 minutes walk) [1]. The path stays on the south side of the burn but just as it approaches some of the old workings a stream is crossed. About another 150 m further on leave the path and head north to join the south bank of Sannox Burn by a fenced-off and flooded shaft (about 50 m north of *No. 1 shaft* marked on Figure 5.1). Here at locality 1 a well exposed barytes vein strikes across the stream (Figure 5.2).

- What physical properties demonstrate that the bright, white mineral is not feldspar [2] [4]?
- Pick up a loose lump of barytes and a same-sized lump of country rock, and hold one in each hand. How do they compare in density? [4]
- What lithological category does the surrounding country rock belong to? [2] [3]
- Are any minerals identifiable in the country rock and if so, what are they? [2] [4]
- What are the physical properties of the country rock? [2] [6]
- What is the lithological name of the country rock? [2] [7]

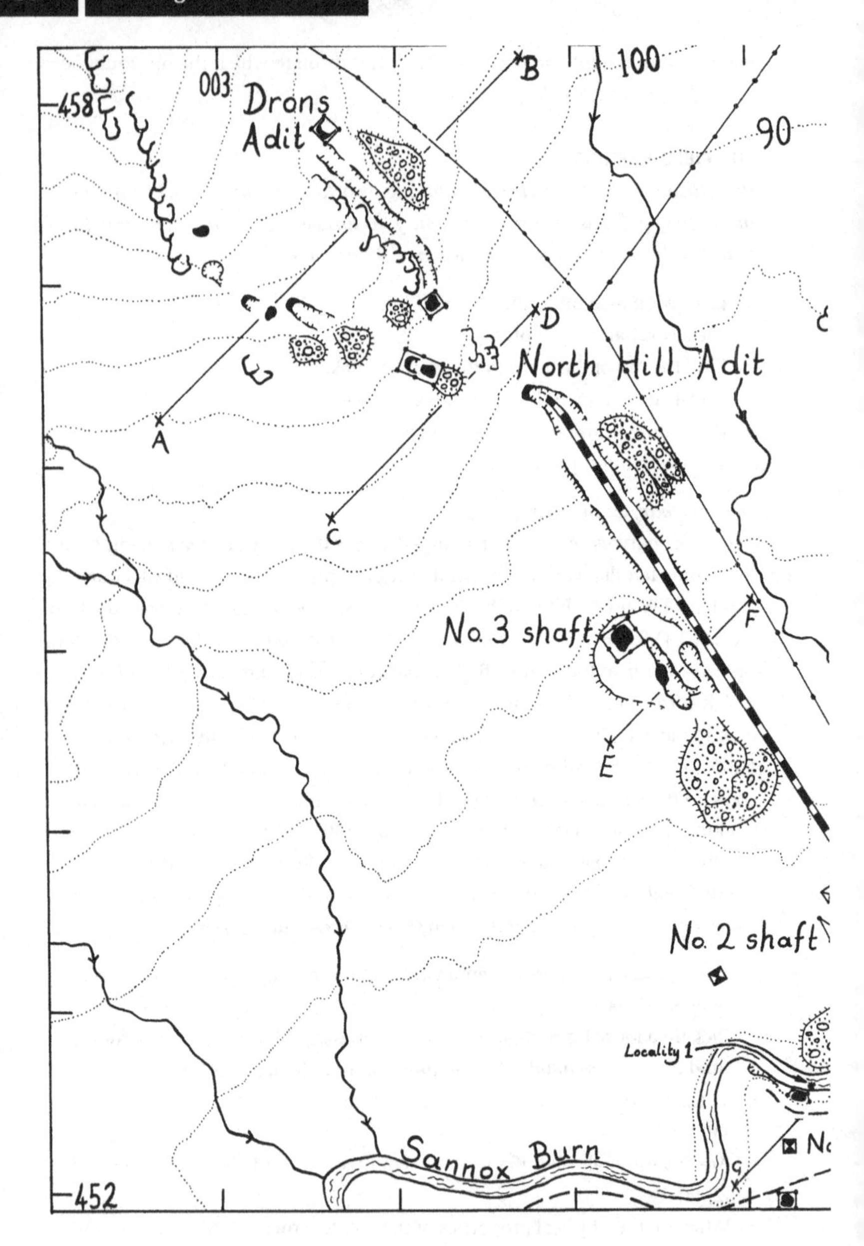

B
100
003
458
Drons
Adit
90
D
North Hill Adit
A
C
F
No. 3 shaft
E
No. 2 shaft
Locality 1
Sannox Burn
452

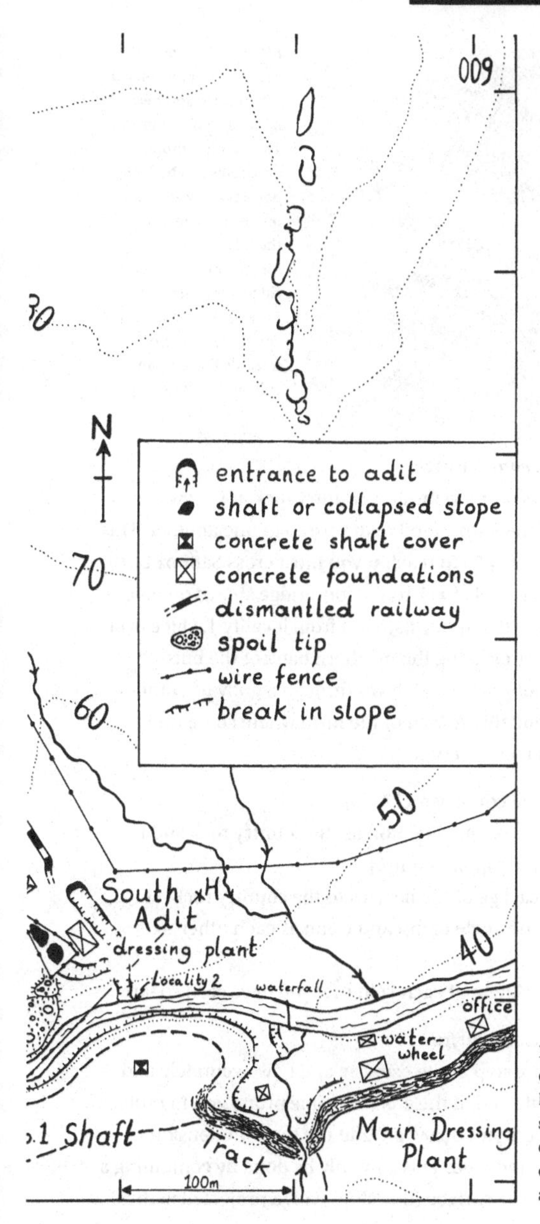

FIGURE 5.1 Exercise base map of the disused barytes mine in Glen Sannox. Foundations for all the old workings are shown along with the names for the original shafts and adits. Lines of cross-section used later in the exercise are indicated by the letters.

FIGURE 5.2 Locality 1 as seen from the fenced shaft on the southern bank of Sannox Burn. The physical properties of barytes can be seen particularly well in these outcrops. Note also how the heated country rock forms a more resistant ridge in the bed of the burn and that the vein can be seen to dip towards the left hand side of the photograph.

Find the dip of the veins to the country rock

After these detailed descriptions of the veins and country rock, it is now important to say something of their structural relationship to one another. This can be done conveniently at locality 2. To reach it you must cross Sannox Burn, and the dryest way to do this is to walk back to the footbridge shown on Figure 1.2 (marked 'f.b.') which is about 400 m downstream from locality 1. Once over the footbridge walk back westwards along the northern bank of the burn to locality 2 [NS 0071 4530]. Carefully scramble down through a gully to examine any of the river cliff section along this stretch of the burn where some thin veins of barytes can be seen in cross-section.

- What is the regional dip of the country rock? [10]
- Do the barytes veins dip in the same direction as the country rock and if not, then what is their dip and dip direction? [10]
- What is the relative geological age of the barytes to the country rock? [14]
- Do the veins all have the same angle of dip and trend as each other? [10] [14]
- What is the approximate thickness of each vein that you have seen so far?

Construct a map of the veins at the surface

This is the most time consuming step of the exercise and how accurately and thoroughly it is done will be reflected in the ensuing mine proposals. In order to know where to mine for the barytes below ground the three-dimensional structure of the veins must be predicted. This can only be done by combining a map of the veins at, or close to, the surface (this step) with a projected vertical slice through the rock (the next step).

FIGURE 5.3 Hunting for barytes veins can be aided by the old mine workings. Here a collapsed stope suggests that a vein is running near the surface. Closer inspection reveals a few patches of white barytes still remaining, indicating the course of the vein. Wherever access is possible and safe the dip and dip direction of the vein contacts should be measured so that the vein can be projected underground.

So for most of the remaining time in this exercise, you should make a map of the barytes veins using Figure 5.1 as a base-map. The entire area south and west of the wire fence must be traversed and any outcrops of barytes marked onto this map **[21]**. In order to construct cross-sections, structural data regarding the dip and dip direction of the veins must be measured and these should also be drawn on Figure 5.1 using the appropriate structural symbols [21]. It is important to mark on the boundaries of the veins whilst mapping **[22]** as this will help predict where they should be found elsewhere on the hill-side (do not try to join up the outcrops of barytes on the map with a ruler!). The thickness of veins should be estimated wherever possible as this information will be needed when trying to decide on a mining strategy for the area. Any other structures such as faults and folds should be noted and the appropriate symbols drawn on the map (these will not have a major influence on this mapping exercise, but it is as well to look out for them in any case) [17] **[23]**. The barytes should be easily spotted at a distance on the hill-side because of its bright white colour. However, where outcrops are scarce the old mine workings might give a clue as to where the veins run (Figure 5.3).

Project the veins below ground on a cross-section

Geological mapping provides a plan view of the veins at or close to the surface. Accompanied by structural information, this allows projection of the veins below ground by constructing a series of cross-sections through the glen. Figure 5.4 shows four parallel lines of section striking north-east to south-west down the northern side of Glen Sannox to the burn (these lines are labelled in

Figure 5.1). As each section is at a progressively lower position on the hillside, veins about 70 m below ground on *section AB* would actually be at the same height above sea level as those at the surface in Sannox Burn (*section GH*).

So construct four cross-sections to illustrate the barytes veins' sub-surface geometry using Figure 5.4 **[24]**. Once this has been done, think about the overall three-dimensional shape of the barytes veins [**25**] before launching into the next step.

Synthesis: mining strategy

The best way of extracting the barytes can now be investigated using your interpretations on the map and cross-sections. A variety of mining methods exist, each dependent on the nature and depth of the *resource* to be mined. The two most appropriate methods for Glen Sannox are those used for extracting rather irregular-shaped mineral bodies. *Deep caved mining* is entirely underground and consists of horizontal tunnels at varying depths (called *levels*), often with removal of the resource from the mine via a vertical lift shaft, or through a tunnel opening onto the hillside (called an *adit*). Mining from the surface can be done using *open-cast extraction*, which essentially produces a large hole in the ground. The waste rock (called *spoil*) is removed to expose the ore body which is then extracted. In both cases the spoil must be transported off-site for dumping.

Consider each of these mining techniques in relation to your sub-surface reconstruction of the barytes veins. To find out which of the two mining methods is likely to be the best

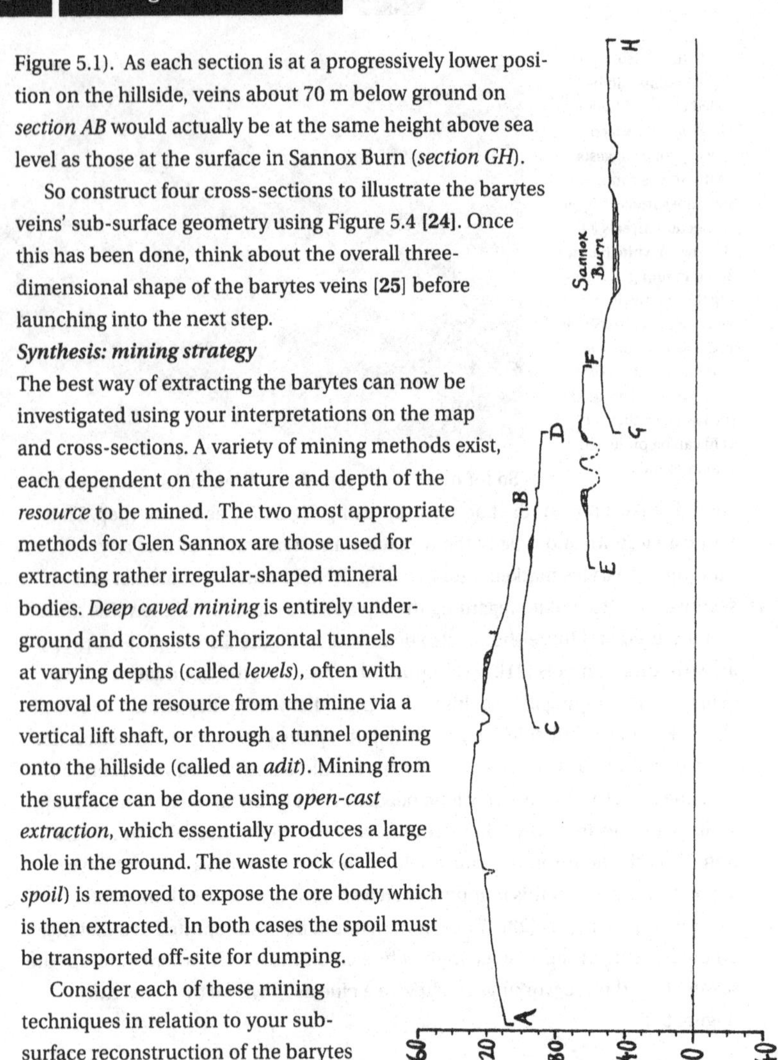

FIGURE 5.4 Four lines of cross-section through Glen Sannox shown at their correct height relative to each other and to sea level. Use [24] to mark onto each section the geological information that you have collected. To help save time, the topography of each section has already been drawn, so you only need to use steps (viii) onwards in [24] for each section.

in Glen Sannox requires calculating the *stripping ratio* (this is the number of tonnes of spoil divided by the number of tonnes of resource mined, i.e. sandstone over barytes). The lower the final figure, the more suitable the technique. Using a ratio allows you to compare directly two mining methods that would each remove a different total volume of rock.

The entire valley does not have to be considered to find the stripping ratio. Imagining a small exploratory open-cast pit or a single level into the hillside will do just as well for the calculation. To help with this, make the volume of rock that has to be removed in each case easier to calculate by using simple dimensions (e.g. by removing a 100 m × 100 m × 100 m cube for open-cast, or a level 2 m × 4 m × 500 m; current mining regulations insist that a level must be at least 4 m wide and 2 m high to let a wagon of ore through without squashing anyone). So draw a square, triangular or rectangular exploratory mining area of your choice for each of the two methods, onto both Figures 5.1 and 5.4. Combining the dimensions of plan view with cross-section will allow you to calculate the total volume of rock that must be removed, the volume within this that will be barytes (take an average thickness for the veins) and consequently the remaining volume that will be sandstone.

- What is the stripping ratio in each case, and so which mining method is best? *NB You need to know that 1 cubic metre of barytes weighs 4.5 tonnes or 4 500 kg (1 tonne = 1000 kg), whereas 1 cubic metre of sandstone weighs 2.6 tonnes or 2600 kg). If your veins are dipping then you may also need to use some trigonometry (sin of an angle = opposite / hypoteneuse).*
- How much money is your preferred mining method likely to make in the exploratory area? *NB Barytes can currently be sold on the export market for approximately £75 per tonne.*
- Could the waste sandstone be used for anything?
- What would be the impact of the chosen mine strategy on the environment and local community?
- The cost of purchasing the old mine, along with additional expenditure such as machinery, the time taken to excavate, transport of the spoil and ore from the island, mine buildings, salaries, accommodation and the clean-up operation after mining has finished, would be *at least* £25 million. Would re-opening the Glen Sannox mine be commercially or environmentally acceptable at present?

Exercise 5: field notes exact location	General location: Date:	data

Exercise 5: field notes exact location	General location: Date:	data

Exercise 5: field notes exact location	General location: Date:	data

Part 3

Applying basic field skills

6 Cuvier's catastrophies * * *

(Corrie, east coast)

PRELIMINARIES

Task To test Baron Cuvier's hypothesis that the sedimentary rock record is a continuous catalogue of sudden environmental catastrophies and extinctions unlike anything seen today.

Logistics *Start Point:* The Corrie Hotel car park [NS 0255 4331] The village of Corrie is situated on the east coast of Arran. The exercise area lies between two prominent ridges on the foreshore. The northernmost of these marks the top of the Devonian succession [NS 0230 4419], whilst the other south of the hotel represents the top of the Carboniferous ([NS 0260 4310]; see Exercise 2). Between these two markers is a sequence of sediments and extrusive igneous rocks, some of which are low-lying outcrops covered at high-tide. Although the exercise can still be attempted at moderately high-tide it should be noted that the amount of exposure will be reduced by about half. Therefore, conclusions drawn during the exercise would be less accurate than at low-tide. *Also, it may be useful to carry a calculator during this exercise.*

Length

3 hr 30 min: 15 min walk to begin first step, Remainder of time to complete the exercise, 5 min walk back to *Start Point*

Field skills required [1] [2] [3] [4] [5] [6] [7] [8] [9] [10] [12] [13] [14] [22]
Background information

The French Baron Leopold Chrétien Fréderic Dagobert Cuvier, known to his friends as 'Georges' (Figure 6.1) and the Englishman William Smith (1769–1839) were the first people to suggest that successive rock strata each contain a distinct fossil assemblage. Without this simple innovation there would have been no geological time scale until the early part of the twentieth century, very few meaningful geological maps at all and arguably no basis for a theory of evolution.

FIGURE 6.1 Baron 'Georges' Cuvier (1769–1832). Cuvier is depicted here about the time of his work on 'catastrophies'

To explain the sudden change seen in the rocks from one fossil assemblage to another, Cuvier hypothesised that the animals and plants were wiped-out by hideous flood events *unlike anything seen today*, with new and more complex species appearing afterwards to replace those that had become extinct. This attractive idea was quickly taken up by many geologists of the time and combined with the commonly held view that the Earth was gradually cooling. Catastrophic environmental change was only to be expected if the climatic belts shifted as polar ice crept towards the equator, and many geologists felt that the rock record showed this to be happening. Essentially, the two concepts together meant that the Earth and the organisms that lived on it were on a one-way ticket through time which would inevitably end with a dead, frozen planet. Not surprisingly this theory was quickly dubbed *catastrophism* and produced no end of debate (and presumably panic) in the scientific and religious community, in many respects causing the foundation of the Geological Society of London.

FIGURE 6.2 Charles Lyell (1797–1875) in 1836, only a couple of years after all three volumes of his *Principles of Geology* had been published.

It was not until a young Scottish landowner, Charles Lyell (Figure 6.2) published the findings of nearly 15 years travelling and geological research that a compelling challenge to catastrophism was offered. Lyell's book, *Principles of Geology* proposed what became and has remained one of the essential ideas behind the geological sciences, and this exercise sets out to explore what it is and why it is so fundamental.

The two contrasting hypotheses of Cuvier and Lyell are examined using Corrie's sedimentary rock record to see how they are supported or contradicted by the field evidence. A Carboniferous succession of terrestrial and marine sediments together with volcanic rocks are investigated in three stratigraphic sections. These sequences illustrate critical geological concepts such as the relationship between stratigraphic thickness and geological time, changing depositional environments, the role of unconformities, and the consequences of sediment compaction after burial. This exercise involves field skills met in Exercises 1, 2, 3 and 5, but the idea of sedimentary logging is also introduced.

THE FIELD EXERCISE

Read through all the instructions thoroughly before beginning work. Annotate maps and figures as required and write your answers to the questions as field notes on the pages provided at the end of the exericse. Remember, there's a 'Help' section at the end in case you get stuck.

Steps needed to achieve the set task

1 Find and examine an example of Cuvier's catastrophies.
2 Consider Lyell's alternative theory to catastrophic events.
3 Assess the evidence for the Earth cooling over time.
4 Understand where Cuvier went wrong.
5 Synthesis.

Find and examine an example of Cuvier's catastrophies

The first step in this exercise is to establish exactly what Cuvier meant by the term 'catastrophy' and what observations he used to support his theory. It is important to realise right at the start that as far as Cuvier was concerned, he based his arguments on field evidence and was simply proposing a hypothesis to explain the rock record. One might be surprised to find an example of a catastrophic flood that wiped-out a whole multitude of organisms in a single stroke on Corrie foreshore. However, there is a good example demonstrating all the apparent hallmarks of Cuvier's theory about 15 minutes walk from the *Start Point.*

To begin the exercise, walk north along the side of the road from the hotel car park for ~250 m until a bend to the left takes it around the harbour at [NS 0244 4355] (Figure 6.3). On the outside of this bend set back from the road

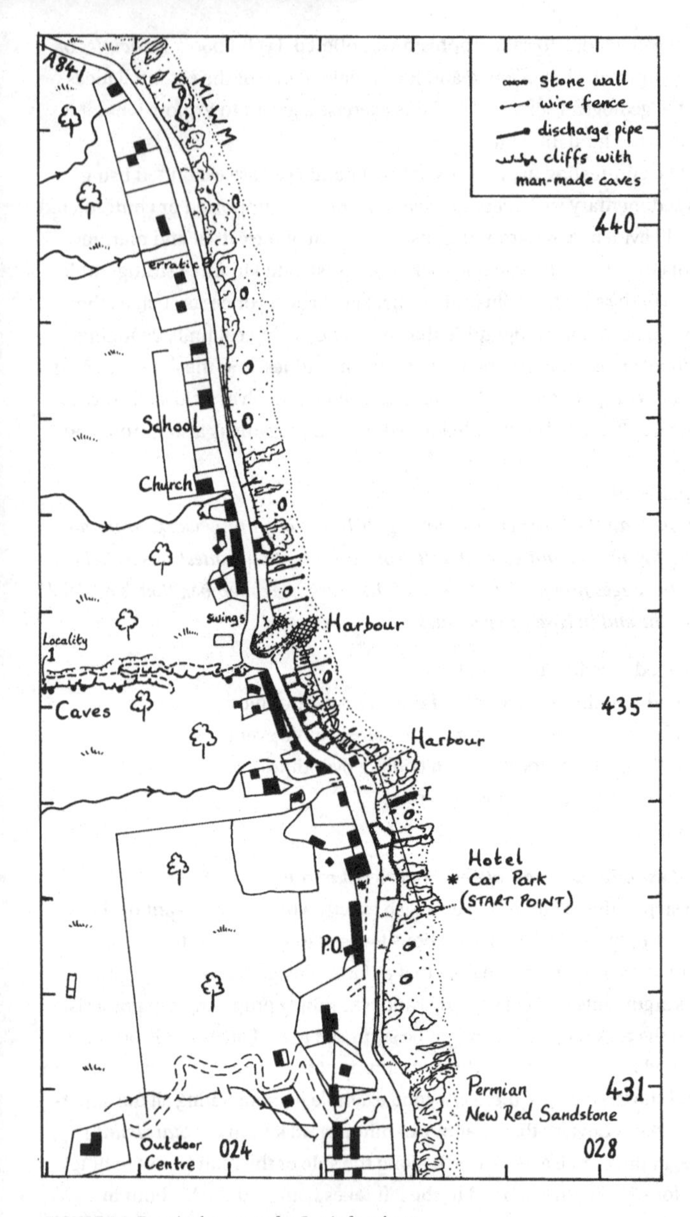

FIGURE 6.3 Exercise base map for Corrie foreshore.

FIGURE 6.4 Looking south at locality 1. The cave walls can be examined to see some of the representative fossils at this locality in cross-section. They can also be seen above in the faulted cave roof. The opposing wall of the cave is just to the right, out of picture and right of this again is the scree slope leading to the overlying beds. *NB Hard hats should be worn.*

by a few metres are two 'caves' into a cliff; actually man-made quarries for limestone. Take the path off the road that leads past these to the right and then loops up an embankment by a brick lime kiln. Keep following this path westwards up the relatively steep and often muddy hillside to the furthest 'cave' marked on the map at [NS 0220 4351]; this is locality 1 (Figure 6.3). In theory, any of the caves in this area would do, but this particular one has access to the units above via a steep scree slope against its west wall. *Do not enter any of the caves as there is a danger of roof collapse.* The field observations here can be done by looking at the walls either side of their entrance. *If you have a hard hat then it should be worn in this area.* Look at the beds that compose the cave walls (Figure 6.4). They consist of a series of repeating thin, red, shaly mudstones and slightly thicker, more prominent, grey limestones. Check that you can recognise these units.

- What fossils can be seen in cross-section on the cave walls? [8]
- Look up at the cave roof, what are the bumps on it? [8]
- Is there any variation in abundance of the fossils throughout the units composing the cave walls?
- How many of the fossils are preserved in their 'life position'? [8]
- What does this suggest about the local water energy and the influx of sediment?
- Are there any other types of fossil here (try looking for bioherms)? [8]
- Why do you think that so many of one type of fossil are here?

Now move to the west side of the cave where a ~10 m slope covered in scree and vegetation leads up to a cliff ~4 m high with good exposure of beds that overlie those of the caves. *Carefully* examine the units that are seen in cross-section on this slope from just above the level of the cave roof up to where it is covered in scree below the cliff exposure. Much of the rock here looks grey/brown because it is covered by a fine layer of 'dirt', but with a little perseverance the underlying lithology can be observed and described.

- What are the physical properties and name of the lithology that composes these more massive beds? [2] [3] [6] [7]
- Would these beds have been deposited from water that had a higher or a lower energy than those in the cave walls? (Use the size grade of the sediment to determine this; the larger the clasts, the more energy the water must have in order to move it). [2] [6]
- How do these units compare with those of the cave walls in terms of their fossil content? [8]
- How might Cuvier have explained the change in fossil content and sediment composition between the cave wall beds and these above?

Now scour the loose scree on the slope in front of the small cliff for fossils. They should be quite common in the red to purple shaly muds and silts that have come from a bed at about shoulder height when standing next to the cliff. Draw sketches of some of the specimens that you find. *Do not collect fossils directly from the cliff as loose sandstone slabs overhang it and they will fall down if undermined.*

- Can any of the fossils be recognised and if so, what are they and what environment would they have lived in? [2] [8]
- Are any the same as in the cave walls? [2] [8]
- How does this appear to favour Cuvier's hypothesis?

In conclusion, it would appear that the field observations *could* support Cuvier's hypothesis. So why was this theory eventually rejected in favour of Lyell's, and just what exactly did he propose? Walk back down the hill to rejoin the road at the harbour and ponder this background to Lyell and his ideas as you go.

Lyell had been trained as a lawyer and he felt strongly that to create a hypothesis, much of which was untestable, was illogical and unscientific. In his view Cuvier may, or may not, have been right but nobody would ever know because Cuvier had maintained that nothing like his sudden catastrophies occurred on the planet today. This seemed to Lyell like just a lot of arm-waving, because no-one would ever be able to prove or disprove a theory that was based on processes that did not now exist. Lyell realised that the only way ever to be scientific about geological history was to tie it to what geologists could see happening around them in the world today: observe what geological processes were forming rocks now, the way in which they operate over time and then apply these to interpret how similar rocks must have formed in the past.

In a letter to his friend Murchison in 1829, Lyell set out the essence of his forthcoming blockbuster book. 'It will endeavour', he wrote, 'to establish the principles of reasoning in the science which are that no causes whatever have from the earliest time to which we can look back, to the present, ever acted but those now acting; and that they never acted with different degrees of energy from that which they now exert'.

Consider Lyell's alternative theory to catastrophic events

Once back on the main road at the harbour, turn north and head for locality 2 about 600 m away. Near the last (or first) house in Corrie, the main road takes a bend to the left and on the outside of it, on the foreshore, is a bench leading to a prominent red ridge of rock striking out to sea [NS 0230 4419]. The large, white clasts in this lithology should be obvious even at a distance. The short sequence from the top of this ridge into the overlying rocks about 75 m south of it is locality 2 (Figure 6.3).

The unit which forms this prominent red ridge is part of the stratigraphic group called the (Upper) Old Red Sandstone and marks the top of the Devonian on Corrie foreshore (see Figure (ii)). It is a coarse conglomerate with lenses of finer, cross-bedded sands. These rocks are separated from a ridge to the south composed of the lowest exposed

FIGURE 6.5 Looking north along the foreshore at locality 2. In the left middle distance can be seen the red ridge marking the top of the Devonian Old Red Sandstone succession and from the middle of the photo to the right foreground is the *'lahar'*.

beds in the Carboniferous by a gully ~0.5 m wide. Yet it is the unit which overlies both these sedimentary ones that provides the most dramatic field evidence at this locality. So continue to walk south over the foreshore and lookout for a sudden change in lithology more or less opposite the last-but-one building in Corrie (Figure 6.5). Once this unit has been found, mark the position of its geological boundary with those beneath accurately onto Figure 6.3 [22], as this will be needed in a later step. Walk through this unit until its upper boundary is reached and again mark this on Figure 6.3 [22] (approximately opposite the next house to the south, *Thirlestane*). Now take a more detailed look at this lithology in an attempt to describe and identify it; make sure to have a good wander around to look at the features present on an outcrop scale too.

- Is there any planar fabric in this unit (e.g. bedding, cleavage)? [9] [15]
- What different types of mineral or lithology can be identified as large clasts (≥10 cm) and what is the approximate size range of each? [2] [4] [5] [7]
- What different types of mineral or lithology can be identified as smaller clasts and/or matrix? [2] [4] [5] [7]
- What are the physical properties of these smaller clasts and matrix? [2] [6]
- Using Figure 6.6, what percentage of the 'whole rock' does each type of mineral or lithic fragment represent? [2]
- To what lithological category does this unit belong? (NB This might seem like a 'trick' question!) [2] [3]

FIGURE 6.6 Imagine this circle printed on the rock, and draw in the minerals and clasts (including lithic fragments) that you see with a hand lens, making sure to show their physical properties such as size and roundness. Then estimate the percentage of each within the circle. For instance, quartz = 60%, plagioclase feldspars = 40%. NB The total must equal 100%!

This is an intriguing rock type and one which from its physical appearance could easily be categorised as yet another 'flood' deposit by supporters of Cuvier. But look again at the evidence you have gathered in answering the questions above.

- The Indonesian name for this 'mystery' rock type is a *lahar*, why do you think that an Indonesian term might be the one given to it?
- What environmental situation today might produce this type of rock, and how does this relate to the last question? *(NB By trying to find present day processes that might have been responsible for forming this rock in the Carboniferous, you are directly using Lyell's principles of geology.)*

Assess the evidence for the Earth cooling over time

By suggesting that geological processes at the present day are fully representative of those that acted in the past, in *kind* and *degree*, Lyell was also attacking the idea of the Earth cooling significantly during its geological history. For instance, he maintained that there were just as many volcanoes erupting lava and releasing heat today as there were millions of years ago and so the centre of the the planet must still be approximately as hot. He had no real explanation for *how* it managed to generate its own heat; this had to wait another eighty years until Rutherford discovered radioactivity. But Lyell's observations were enough to demonstrate that the Earth, rather than being on a one-way ticket through time, must actually have a certain dynamic stability; retaining a certain *uniformitarianism*.

This step examines evidence for a dynamic stability or cyclicity in the Earth's processes rather than a directional force. So walk south along the road from the *lahar* and return to the harbour at [NS 0244 4355]. From here to just south of the *Start Point* on the foreshore at [NS 0260 4329] is a sequence of Carboniferous sedimentary beds (Figure 6.7). Every individual exposed bed along the beach here is shown in a sketch cross-section (Figure 6.8). So start at the south wall of the harbour and walk over the beds on an approximate bearing of ~166°. Use Figure 6.8 to build up a *log* of

FIGURE 6.7 Looking south from the harbour over the units shown in the sketch cross-section of Figure 6.8. Notice the prominent hut on the sky-line which is half-way through the succession studied here.

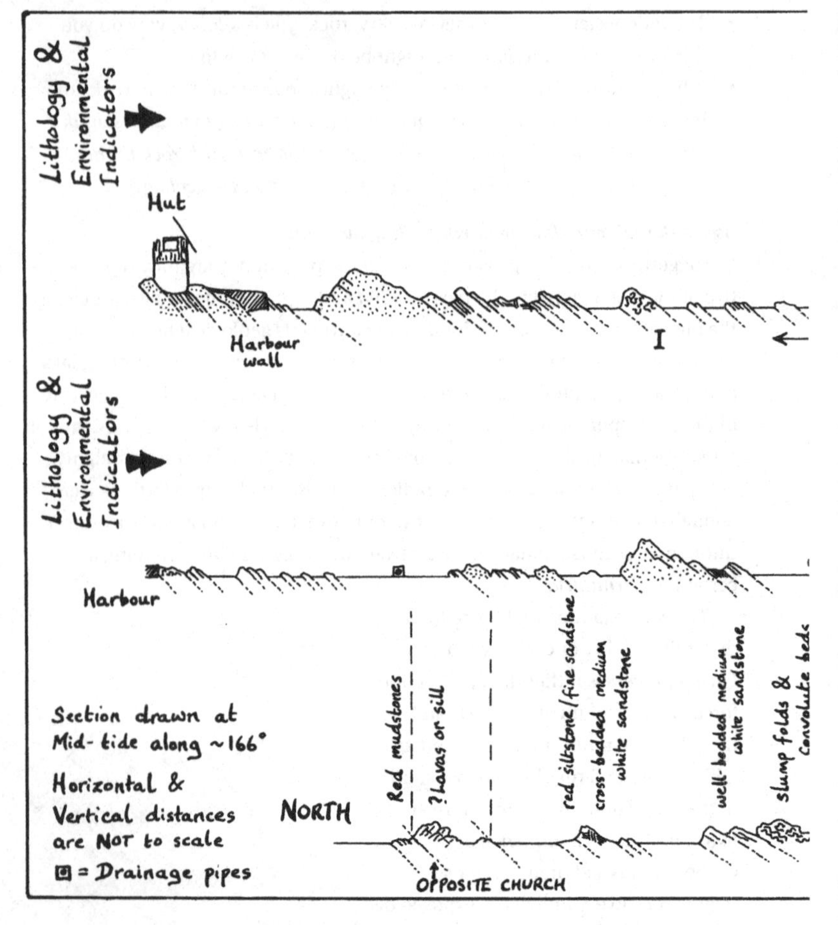

FIGURE 6.8 Sketch cross-section along Corrie foreshore from north to south showing all beds that outcrop. Note that the beds seen at the caves in the first step actually strike down onto the foreshore and have been quarried away to produce the harbour. Their relative position in the stratigraphic succession is shown so that this information can be combined with evidence collected on the foreshore. Use this section to record detailed lithological descriptions ([2] [3] [4] [6] [7]) and environmental indicators ([8] and [9]) in each bed. Examples of how to record the information are given for units north of the harbour. Colour-in the lithologies on the section according to the size grade of the sediment [6].

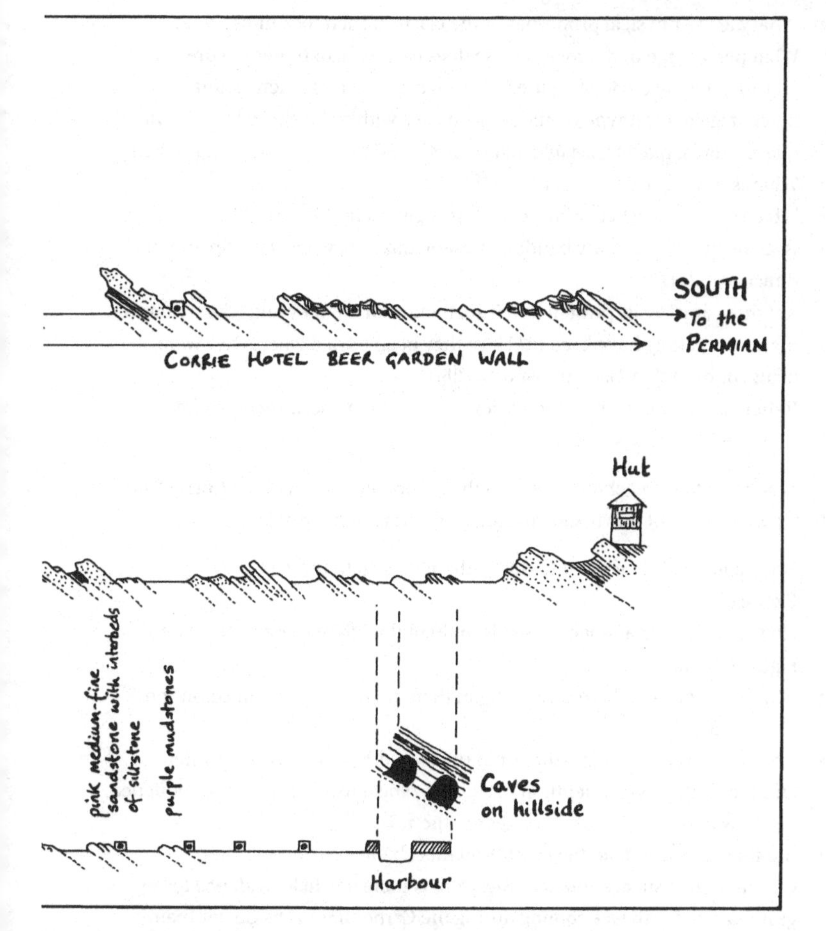

the sedimentary succession, which is a common method geologists use to record how the depositional environment changed through time. Be sure to look out for any environmental indicators that might be present in the rocks, such as fossils (particularly *terrestrial* fossils) [8] or specific sedimentary bedforms [9] and make a note of these on Figure 6.8 as well. Use the questioning procedure below to identify the lithologies in each unit of the sequence.

- To which lithological category do the rocks belong? [2] [3]
- What different types of mineral can be identified as clasts and what is the approximate size range of each? [2] [4] [6]

- What are the physical properties of the clasts in each rock unit? [2] [6]
- What percentage of the rock sample does each type of mineral represent? [2] (Imagine the circle of Figure 6.6 printed on the rock, then estimate the percentage of each type of mineral and clast within the circle. For instance, quartz = 60%, plagioclase feldspars = 40%. NB The total must equal 100%!)
- What is the name of the rock? [2] [7]
- What sedimentary bed-forms can be recognised in the unit? [9]
- Is there any sedimentary bedding present and if so, what is its dip and dip direction? [2] [10]
- Are there any fossils present and if so, then what are they and what environment would they have lived in? Have they been buried where they were living, or have they been transported?[8]
- If there are no fossils present, then why; might it be because of where the sediment was deposited?

When you have reached the end of the section just south of the Corrie Hotel beer garden wall, sit down and consider the completed log of Figure 6.8.

- How many beds had plant fossils in them (highlight these in colour on Figure 6.8)? [8]
- How many beds had marine fossils in them (highlight these in colour on Figure 6.8)? [8]
- How many beds had cross-bedding in them (highlight these in colour on Figure 6.8)? [9]
- Approximately how many times did the local depositional environment, and the relative water depth in particular, repeat itself through the sequence (i.e. how many sedimentary cycles are there)?
- Are there any trends to the environmental change up through the Carboniferous succession, and therefore is there any field evidence to suggest that the Earth was cooling during the Carboniferous as Cuvier maintained?

The cave sequence of the first step was simply an isolated example from one of these cycles and you should have found a marine limestone on the foreshore with a similar fauna to that in the scree overlying the caves. The appearance of the same fauna, whether marine or terrestrial, in repeated cycles suggests that they are not wiped-out across the globe in a 'Cuvier catastrophy' when they disappear from the rocks on Corrie foreshore, but are simply killed locally and recolonise from elsewhere when favourable environmental conditions return. The cyclicity through Carboniferous sediments at Corrie also demonstrates Lyell's idea of uniformitarianism rather than a continual cooling of the Earth.

Understand where Cuvier went wrong

If Cuvier's hypothesis was based, as he stated, on observations of the sedimentary rock record then where did he go wrong in the interpretation of it? In essence, the pivotal error was that he failed to grasp the importance of the relationship between stratigraphic thickness and time. Cuvier saw thick sedimentary formations containing the same fossil fauna all the way through them, overlain by another thick formation with a totally different fossil assemblage. The change from one to another was literally across a line that you could put your finger on in the field. Cuvier's assumption was that *the thickness of rock that he saw was directly proportional to time.* His hypothesis provided no means of testing this using anything happening at the present day. However, in contrast, Lyell's ideas provided an opportunity to calculate the approximate time that must have been needed to deposit particular sedimentary beds in the rock record based on observations of their present day equivalents. What this actually shows is that *the stratigraphic thickness of a sedimentary bed is not proportional to the length of time needed to deposit it.*

This can be tested easily on Corrie foreshore by comparing the time taken to deposit two different beds, each of a different thickness. The first is the *lahar* investigated in step two (which contains no internal 'layering' of any kind so it can be assumed to have been deposited as one single unit). The second is the bed marked **I** on Figures 6.3 and 6.8. Consider again the depositional environments of both.

Unfortunately, the rocks of Corrie foreshore are not exposed in perfect cross-section to allow an immediate comparison of the two units. Instead a calculation must be made which corrects for the influence of later tectonic movement in the area which has left the rocks of Corrie all dipping at a low angle to the horizontal. So just measuring the distance on the ground through each unit will not give a true value of their stratigraphic *thickness* but a greatly exaggerated figure instead. However, it can be found using a simple trigonometric equation if two other variables are known: the horizontal thickness of each unit (h) and the regional angle of dip (α) (Figure 6.9). The upper and lower geological boundaries of the *lahar* should have already been marked on in step two, whereas the outcrop of **Bed I** is shown on the map and so the boundaries can be drawn on now [22].

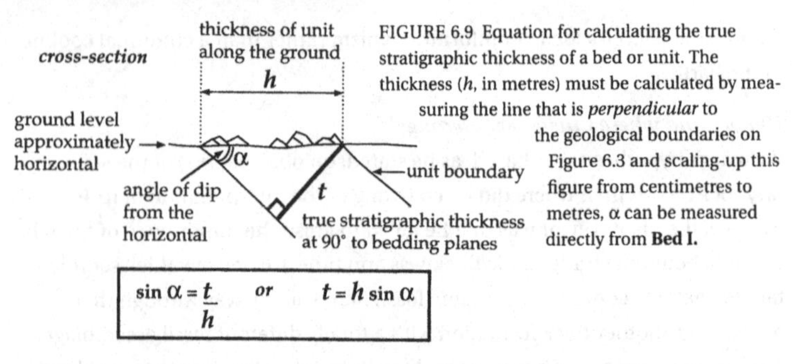

FIGURE 6.9 Equation for calculating the true stratigraphic thickness of a bed or unit. The thickness (h, in metres) must be calculated by measuring the line that is *perpendicular* to the geological boundaries on Figure 6.3 and scaling-up this figure from centimetres to metres, α can be measured directly from **Bed I.**

- What is the scale factor for Figure 6.3? (Use the 100 m grid lines to find the scale factor. For instance, if 1.5 cm on the map represented 100 m on the ground, then the *scale factor* would be 100/1.5 = 66.66).
- What is the measured thickness on the map (h) for the two units (in centimetres)?
- What is the actual value of h (in metres) for each unit? (If h were measured as 3 cm on the map, then using the scale factor above, its real value would be $3 \times 66.66 = 200$ m).
- What is the average angle of dip and dip direction of this sequence? [10] (A regional angle of dip for the rocks in the sequence (α) must use a sedimentary bed that can be assumed to have been deposited horizontally, such as **Bed I**, rather conveniently).
- From your values of h and α, and using Figure 6.9, what is the true stratigraphic thickness of the *lahar*?
- Approximately how long would it have taken to deposit the *lahar* if it was all from one event – minutes, hours, years, thousands of years or millions of years?
- From your values of h and α, and using Figure 6.9, what is the true stratigraphic thickness of **Bed I**?
- Would **Bed I** have been deposited faster or slower than the *lahar*?
- Does this agree with Cuvier or Lyell?

Synthesis

Ultimately the thickness of the rock record will depend on a variety of interacting processes. The rate of sedimentation at a point in time is dependent upon the amount of sediment coming in and the space available in which to deposit it. The former will depend on the energy of the depositional environment and the source for the sediment itself. The latter factor is essentially reliant on how quickly the area is subsiding, either tectonically or by *compaction* of the soft sediment as it is buried. Computer programs now exist which can model *subsidence* based on these variables, but geologists have been doing this for years intuitively in the field. If subsidence, sedimentation rate and lithology are constant in time then *compaction* through the sequence would describe a smooth curve with depth. In reality this situation rarely develops and one or more of the variables change. Think about this with respect to the Corrie foreshore field evidence.

- Is the sedimentation rate constant throughout the sequence of Corrie foreshore?
- Are there any breaks of deposition in the sequence and why might these apparently show *catastrophies*? [9]
- In what ways might compaction cause the rock record to apparently show *catastrophies*?
- Can Lyell's ideas alone explain the entire geological rock record, or are there any instances when a Cuvier-style catastrophy might have occurred? (There is currently a movement for 'neo-catastrophism'; for instance a meteor shower causing the extinction of the dinosaurs ... what would Lyell think?)

Help!

1 Figure 6.10 (a) A Carboniferous rugose coral buried where it was living, surrounded by shelly debris. (b) Carboniferous productiid brachiopods still with both valves intact and in position are likely to have been buried alive by a 'pulse' of sediment. Both these localities are in the same bed on the north shore of Arran, but the equivalent stratigraphic unit is seen at Corrie.

(a)

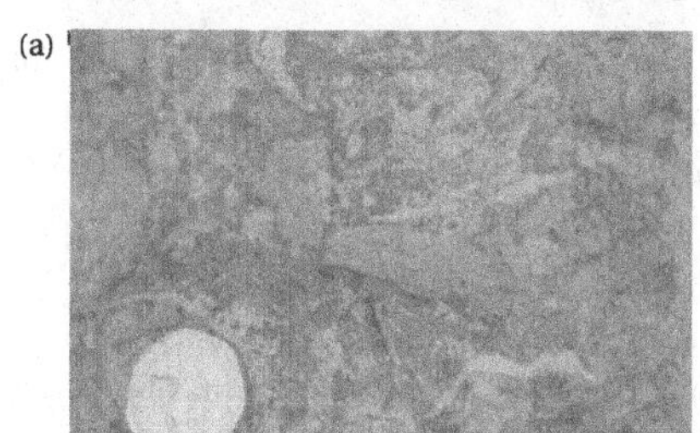

(b)

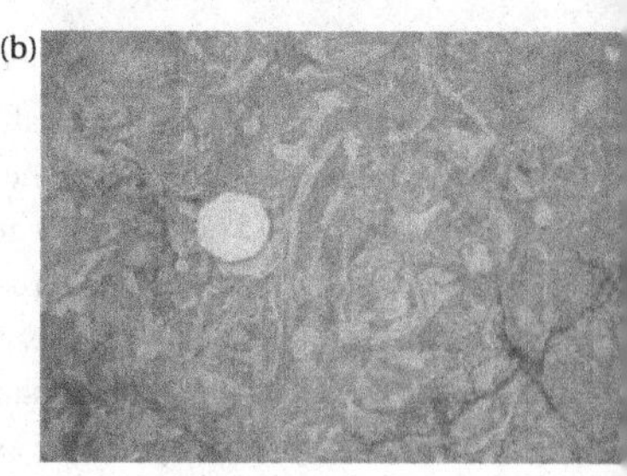

2 If there is a problem in identifying the 'mystery unit' at locality 2 and an environment in which it might have formed, then move on to look at the overlying unit and try identifying this first instead. It stretches south to the school and both the lithology and environment in which it might have formed are easy to identify. The importance of it is that there are lumps of this easy-to-identify rock as *clasts* in the 'mystery' unit underneath. This links the two to the same palaeo-environment.

3 The deposition of transported products is rarely constant, but more periodic in nature. This produces 'packets' of deposited sediment that are preserved in the geological record as distinct 'beds'. The time interval between their accumulation can be so short that it is indistinguishable in the rock record, or it may be so long that the beds beneath have been eroded or deformed and an *angular unconformity* produced [9].

4 Here are some useful values when converting the apparent stratigraphic thickness of a unit over nearly horizontal ground (h) to its true stratigraphic thickness (t) using the average, or regional, dip of beds in the unit (α) if a calculator is not handy: sin 25° = 0.42; sin 30° = 0.5; sin 33° = 0.54; sin 35° = 0.57; sin 40° = 0.64. For example, rocks with a regional dip of 30° would give sin 30° = 0.5. So in the equation $t = h \sin \alpha$, the true stratigraphic thickness of each unit would actually be half its apparent horizontal thickness across the foreshore.

5 Figure 6.11 (a) A bivalve being buried in a matter of minutes by sand in a stream on the foreshore at Blackwaterfoot today. (b) Not enough sediment is

(a)

(b)

being brought in to cover the bivalve sufficiently and the sand deposited inside the valve is soon washed away. The shell is then picked up by the water current and transported further downstream. Shells like this undergoing repeated cycles of transport–deposition–erosion in stream or tidal water currents are gradually ground down over time. Notice that the bivalve shown has already lost its other valve and could remain on the beach being broken into smaller and smaller fragments by the waves until it is the same size as the quartz grains that surround it.

6 The amount of compaction that a soft sediment will undergo can be thought of as being a function of how much fluid-filled 'space' there is in it (its *porosity*) and the nature of the sedimentary grains of which it is composed (i.e. what minerals they are and how large). For instance, a 'typical' arenite will be composed almost entirely of quartz clasts. If it were deposited in water then each clast might originally have been enirely surrounded by a film of this water. Compaction would squeeze this out until the clasts touch and rest against each other; the hardness of quartz probably preventing it progressing much further. So these types of sand*stone* still tend to be *porous* and will still let fluids pass through them; they are *permeable* (oil from the North Sea is pumped out of such sandstones). Mudstones or clays are the complete opposite. In these cases not only can the water be squeezed out, but the mica flakes or clay minerals themselves are 'malleable' and can be pressed flat. Lots of incredibly small pore spaces might still be left but they are isolated and so make the rock impermeable.

Exercise 6: field notes exact location	General location: Date:	data

Exercise 6: field notes exact location	General location: Date:	data

Exercise 6: field notes exact location	General location: Date:	data

Exercise 6: field notes exact location	General location: Date:	data

Exercise 6: field notes exact location	General location: Date:	data

7 Prof. Speyside's theory ***

(Glen Rosa, east Arran)

PRELIMINARIES

Task A new theory has challenged the accepted view that Arran was glaciated during the Pleistocene. The Geological Society is in uproar. Your task is to assess the field evidence and decide what to report at the next meeting.

Logistics *Start Point:* Lay-by opposite the track to Glen Rosa [NS 0041 3678]

The entrance to Glen Rosa lies to the north-west of Brodick on the east coast of Arran. The *Start Point* is only a ten to fifteen minute walk from the centre of town and is located close to the junction of the main coastal road around Arran and 'the String' which connects east to west directly across the centre of the island. Access to the exercise area is via the signposted track into Glen Rosa which can be found on the opposite side of the road from the *Start Point.*

NB This exercise requires fair weather conditions and good visibility as it involves walking into the granite mountains to examine their geomorphology. A steep path is followed up the side of the glen over rough or boggy ground and this might prove a potential hazard to those unused to hiking in upland areas. However, on a clear day the view is worth the climb!

Length

4 hours: 30 min walk to begin first step, 2 hr 45 min to complete the exercise, 45 min walk back to *Start Point*

Field skills required [1] [2] [3] [4] [5] [6] [7] [12] [13]

Background information

The Quaternary Era can be split into two Epochs; the oldest is the Pleistocene (which means 'most recent' and extends from 1 600 000 to 10 000 years ago) and the youngest is the Epoch we are still in today, the Holocene (meaning 'wholly recent', from 10 000 years ago to the present). The Pleistocene was introduced by Charles Lyell in 1839 and applied to deposits containing more than 70% of the mollusc species that still live today. In the late nineteenth century the Swiss geologist Agassiz maintained that extensive glaciation had occurred during the Pleistocene. Consequently, it soon became dubbed the 'Glacial Epoch', or 'Ice Age'.

At a special meeting of the Geological Society in London the highly respected geologist Prof. Stanley Speyside recently challenged the generally accepted view that Arran was covered by an ice sheet during glacial episodes in the Pleistocene Epoch. Speyside reinterpreted the field evidence in northern Arran to form a new theory which can be condensed into six essential points (given below in chronological order).

1 Arran was uplifted during the Late Tertiary, accompanied by subsidence in Kilbrannan Sound and the Firth of Clyde.
2 Earlier Tertiary and Mesozoic cover was removed by sub-aerial erosion almost down to present-day levels by the start of the Pleistocene.
3 Then either subsidence and/or eustatic sea-level change caused a relative sea-level rise on Arran to a stand-point three hundred metres above current sea-level (the Ordinance Datum, or OD). A period of stasis at this higher stand rounded-off the topography as a marine erosional platform.
4 The relative sea-level rise then continued to a final high-stand of ~600 m OD. Only the highest northern granite peaks survived sub-aerially exposed as islands. Drowned former river valleys were eroded by wave action, rounding and widening them into a network of inlets and bays.
5 Uplift or eustatic sea-level fall then caused the relative sea-level to drop again steadily almost to OD, smoothing out some of the previous erosional features associated with the initial rise. Sediments associated with this event are poorly sorted beach deposits which now drape the topography, deposited across the hillsides as the sea retreated – these are called *drift* deposits.
6 The closing stages of retreat saw two short-term sea-level stands which cut the coastal notches or *raised beaches* now seen around the coast of Arran, before the sea finally retreated to the present OD.

Speyside's theory received a harsh reception by those who work on Quaternary geology elsewhere in Scotland as it attempted to overturn the established view. However, some felt that the theory had something new to offer and presented a fresh approach to interpreting Scotland's geomorphology. Those who oppose Prof. Speyside's hypothesis suggest that a major upper Devensian Stage ice sheet (26 000 to 10 000 years ago) covered Britain south to about Birmingham. This ice sheet began to melt about 18 000 years ago and it is likely that it had retreated north of Arran by about 13 500 years ago. A sharp deterioration in climate took place between ~11 000 and 10 000 years ago and this caused ice to return to much of western Scotland (this is known as the 'Loch Lomond Readvance'). Each successive advance and retreat of ice would partially or wholly obliterate the features left by the previous one and so much of the evidence on Arran interpreted as glacial in origin has been attributed to the final ice readvance of 11 000 to 10 000 years ago.

The meeting began to disintegrate and the President had to intervene. Since the paper Speyside presented had caused such controversy, it was decided to reconvene at an emergency meeting two weeks later. In the meantime, you decided to check the evidence for yourself and hastily booked a ferry ticket to Arran. Glen Rosa reaches to the very heart of the northern mountains and so should be the ideal area in which to test Speyside's hypothesis in the field.

This exercise draws primarily on field skills learnt in Exercise 1, so it would be useful to have done this exercise beforehand. However, it also requires use of some skills from Exercises 2 and 3 as well. It investigates the opposing geomorphological landforms that might be expected in coastal and glacial environments. Mechanisms for eustatic and apparent sea-level rise are examined with the evidence that would be preserved in the rock record. In addition, this exercise demonstrates the problems sometimes encountered when the same field evidence is interpreted in two different ways and illustrates the pitfalls of accepting a scientific theory without adequate investigation.

THE FIELD EXERCISE

Read through all the instructions thoroughly before beginning work. Annotate maps and figures as required and write your answers to the questions as field notes on the pages provided at the end of the exericse. Remember, there's a 'Help!' section at the end in case you get stuck.

Steps needed to achieve the set task

1 Examine and describe an example of Speyside's *beach deposits.*

2 Identify and assess Speyside's *marine erosional platform.*

3 Re-assess evidence of former *granite islands* at 600 m.

4 Synthesis.

***Examine and describe an example of Speyside's* beach deposits**

Stage 5 of Speyside's theory maintains that Arran's northern mountains were covered in a drape of beach deposits from the retreating sea. Opponents would argue that these *drift* deposits are from the retreat of ice instead. Since in both cases these sediments represent the last event to have taken place, then they should be the easiest to examine (look again at Exercise 1 if you are unsure of the physical properties of beach sediments). So the first step is to examine the physical properties of these beach or drift deposits to see if they provide any clues as to their true origin.

To begin the exercise, cross the 'String' from the *Start Point* [NS 0041 3678] and take the single track road signposted to Glen Rosa (Figure 7.1). Follow this over the bridge, past the farm and cottage to a gate and campsite [NS 0015 3759]. Go through it and on the other side the road becomes a dirt track and skirts around the forest on the northern flank of Am Binnein through two more gates to a sheep dip [NR 9937 3791] (incidentally, this gate is in the bottom right-hand corner of Figure 7.2). The walk to this gate from the *Start Point* should take about thirty minutes or so. From here the route continues as a good footpath along the south-west bank of Glenrosa Water.

A small stream down from Am Binnein to the left crosses the footpath only about 40 m further on from the sheep dip. The next stream to cross the path is in another 60 m and this contains an excellent cutting through the unconsolidated *drift*. So leave the path and scramble ~50 m uphill from the track along

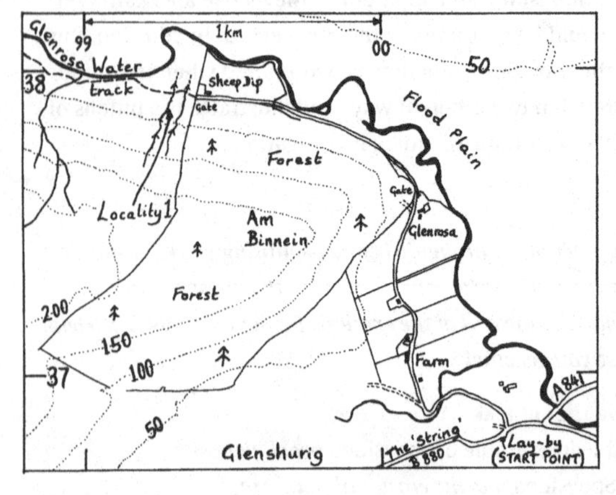

FIGURE 7.1 Location map for lower Glen Rosa, showing the *Start Point* and the position of locality 1. The 'U'-shaped meander in Glenrosa Water just north of locality 1 can be found easily in the bottom right-hand corner of Figure 7.2.

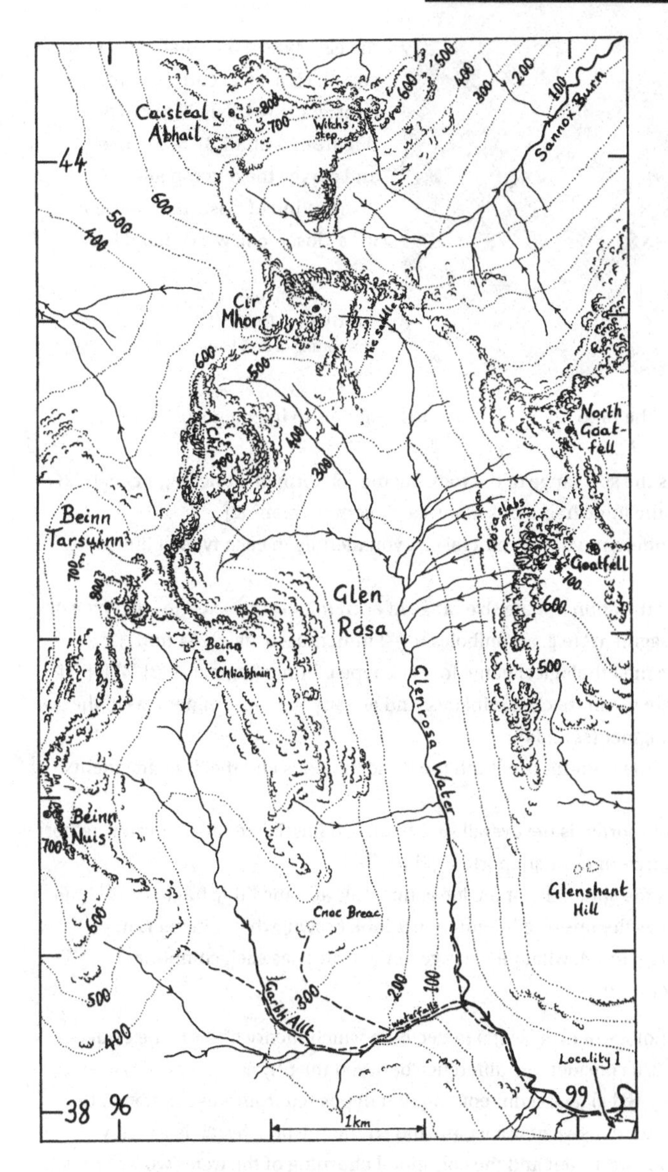

FIGURE 7.2 Exercise base map for Glen Rosa. The major granite peaks surrounding the glen are marked, as well as Cnoc Breac and the path leading to it via Garbh Allt. NB This map is on a much smaller scale than Figure 7.1.

the east side of the small burn until you reach a small cutting in the bank; locality 1 [NR 9930 3789] (Figures 7.1 and 7.3). Stand back and survey the outcrop and the distribution of clasts then move in for a closer look with a hand lens.

FIGURE 7.3 Locality 1, looking eastwards at the cutting of *drift* in the bank of the small burn.

- What is the ratio (in %) of individual minerals to fragments of rock (e.g. 30%:70%)? [2]
- What is the size range of the rock fragments (termed 'lithic fragments')? [2]
- Are all the lithic fragments of the same composition? [2]
- What different types of mineral can you identify in each type of lithic fragment? [2] [4]
- What is the approximate size range of each mineral 'species' in each type of lithic fragment? (e.g. amphiboles ≈ 1–2 mm, muscovite ≈ 2–4 mm) [2]
- What is the lithological name for each type of lithic fragment? [2] [4] [5] [7]
- What are the loose clast minerals, and are any the same mineral as in the lithic fragments? [2] [4]
- What is the roundness of both the loose minerals and the lithic fragments? [2] [6]
- How well sorted is the overall sediment; is it possible to distinguish whether it is matrix- or clast-supported? [2] [6]
- What is the sphericity of the loose minerals and the lithic fragments? [2] [6]
- Why does the finest sediment matrix look brown; what mineral is it?
- Using Figure 7.4, what percentage of the drift does each component represent? [2]

The lithology of drift is vital in deciding which theory best fits the evidence and there is a considerable difference between the physical properties of unconsolidated glacial sediment and that deposited from wave-action at the coast. The zone between wave base and strandline on a beach is a highly energetic environment and the continued churning of the water works the sediment over again and again, rounding and sorting the clasts. Finer debris tends to be picked up during this churning and held in suspension until

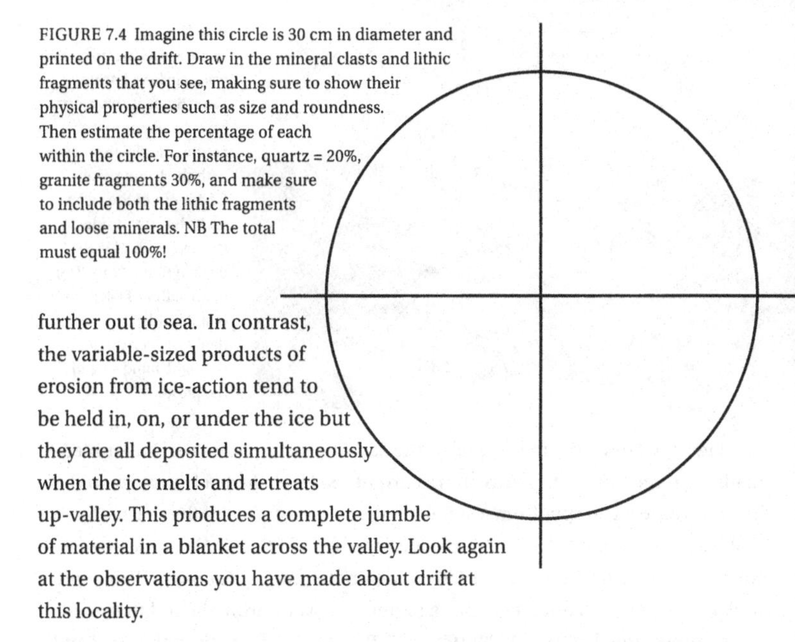

FIGURE 7.4 Imagine this circle is 30 cm in diameter and printed on the drift. Draw in the mineral clasts and lithic fragments that you see, making sure to show their physical properties such as size and roundness. Then estimate the percentage of each within the circle. For instance, quartz = 20%, granite fragments 30%, and make sure to include both the lithic fragments and loose minerals. NB The total must equal 100%!

further out to sea. In contrast, the variable-sized products of erosion from ice-action tend to be held in, on, or under the ice but they are all deposited simultaneously when the ice melts and retreats up-valley. This produces a complete jumble of material in a blanket across the valley. Look again at the observations you have made about drift at this locality.

- Locate the exercise area on Figure (i). The bedrock in this part of Glen Rosa is Devonian Old Red Sandstone (Figure (ii)), so are the lithic fragments and loose clast minerals weathered and eroded bedrock?
- How can the presence of other clasts be explained, and where must they have come from?
- What is it about the orientation and roundness of the large lithic fragments that casts doubt upon them having been continually reworked on a beach?
- What is it about the ratio of matrix to the larger clasts (the sorting and maturity of the sediment; [6]) that suggests a beach is the wrong interpretation of depositional environment?

Identify and assess Speyside's **marine erosional platform**

With doubts beginning to be cast over Speyside's theory, the next step is to look at the wider geomorphology of Glen Rosa to see whether glacial or coastal processes are responsible for the form of the valley-sides and surrounding granite peaks. To get a decent view of these features you will have to walk up the side of the glen. This may prove a little arduous, but if the visibility is good then the view is usually reward in itself.

FIGURE 7.5
View northwest from the track close to Locality 1. Cnoc Breac is the sunny, rounded summit in the centre middle distance with the tributary of Garbh Allt in the valley on its left-hand side. In the background is the peak of Beinn Nuis. The main valley of Glen Rosa is still hidden by the flanks of Glenshant on the right-hand side of the photo.

Stage 3 of Speyside's theory makes the case for a prominent marine-eroded platform three hundred metres above current sea level. Such a feature should be relatively easy to spot from a suitable vantage point. One of the few paths that leads up the side of Glen Rosa follows the burn of Garbh Allt, which heads west towards Beinn Nuis (Figure 7.5). So rejoin the track at Glenrosa Water and walk the 800 m or so until it crosses a footbridge over Garbh Allt and then splits on the other side [NR 9835 3870] (Figure 7.2). Turn left, take the path uphill and start climbing. There are some good exposures of granite down in the stream bed and it may be that some of the drift has become exposed on the stream bank. If so, it may be worth stopping to see if its composition is any different from locality 1. Also, whilst walking, consider the sporadic covering of large boulders across the landscape. Stop to look at one and identify the lithology [2] [3] [4] [5].

- How can the distribution of these boulder *erratics* be explained in relation to either the 'Speyside' or the 'glacial' theory?

About 150 m of height needs to be gained before a fork in the path is reached at [NR 9775 3851]. Take the right-hand branch, which leaves the stream and strikes towards the summit of Cnoc Breac to the north-west. The view should open out, particularly towards Goatfell on the other side of the glen to the north-east, and also to the south. A well-earned rest from the ascent can be taken anywhere along the path here to consider Speyside's theory once again.

- Is a flattened 'platform' visible anywhere in the landscape of this area?
- At approximately what height does it lie? Identify the height from Figure 7.2 and then colour the area on it represented by the platform.

Both ice-action and marine erosion can produce levelled topography. So it is not so much the platform itself which gives clues as to its origin, but the junction of it with the higher peaks. If this were an ancient coastline as Speyside proposes, then it is reasonable to expect that coastal features would have had time to develop along this line such as foreshore, cliffs with perhaps sea-caves, natural arches and stacks. Draw this proposed coastline on Figure 7.2 using the crags as a guide to where it would go. Now look around you at this junction between platform and peaks.

- What can you see at this boundary and does it support or contradict Speyside's thinking here?
- If the 'glacial theory' is favoured, then you must explain how ice might be able to level-off the topography; what form must it have taken?

Re-assess evidence of former* granite islands *at 600 m

If the weather is good and there is time and energy left, it is worth gaining a little more height before attempting the next step. However, if at all unsure then opt for the next best thing and stay at this locality. The geomorphology of the Glen Rosa area to the east and south should still be mostly visible but the view to the north and north-west will be a little obscured. If this second option is taken then miss out the next paragraph and go straight on to the following one.

To gain a better all-round view, carry on along the footpath about another 500 m to the small summit of Cnoc Breac [NR 9727 3918]. Look at the view around you to the north. This takes in most of the granite mountains on Arran from Goatfell [NR 9910 4152] to Beinn Nuis [NR 9556 3990], including a fine view across the head of Glen Rosa towards Cir Mhor [NR 9733 4308] and the Witch's Step beyond.

Highlight the 600 m contour on Figure 7.2 with a coloured pencil. This should represent Speyside's highest stand for sea-level. Draw a panorama field sketch of the landforms on and above the platform, and use the questions below to help label the geomorphological features on it [13].

- Are there any typical coastal erosional features at this height?
- What form must the weathering have taken to produce the shape and texture of the peaks above 600 m?

- What does this weathering suggest about the temperature of the Pleistocene terrestrial climate?
- There are clearly 'embayments' and 'headlands' as Speyside suggests, but why do they not look like they have been carved out by the sea?

There may be little evidence for coastal features at this height, but that does not necessarily mean that the 'glacial hypothesis' has any better claim to be correct unless evidence of typical glacial features can be recognised. Classic features in such environments include steep-sided, 'frost-shattered' peaks that fall away sharply into 'bowl'-shaped hollows. These features are particularly characteristic of accumulation areas for ice, known as *corries*. Ice building up against a peak can eventually start to 'slump' under its own weight to flow downhill as a glacier. This 'slumping' away from the peak tends to have a rotational, 'scooping' effect to begin with, and this carves out the corrie. Corries backing onto one another tend to form very narrow, sharp ridges joining the peaks together, known as *arrêtes*. Once the ice starts to flow it does not follow a sharp, incised channel like a river because generally there is simply too much ice. It may exploit a channel that already exists, but either way it carves the bedrock into ideally a 'U'-shape in cross-section.

- Are any of these glacial features visible in the Glen Rosa area? If so, label and colour them on Figure 7.2 and your field sketch.
- What shape is Glen Rosa in cross-section and can this really be explained by marine erosion; why should topography fall away so sharply from the platform into the valley?
- If, for instance, the mountain summits were poking through the ice cover to be weathered, then what would have been the thickness of ice on the platform at that time?
- What would have been the thickness of ice in the valley?
- From the geomorphological evidence, in which directions did ice flow? (Draw a series of arrows to indicate this on Figure 7.2.)
- How does the present-day drainage pattern relate to the direction of ice-flow?

Synthesis

Return back down the path to the bridge at Glenrosa Water whilst contemplating the mechanisms that could cause an apparent rise in sea-level. A rise of 350 m and especially one of 600 m are orders of magnitude too high for the volume of water in the oceans to be increased from simply melting polar ice (a 'eustatic' rise). Not only would this require runaway global

Evidence in support of Speyside's theory	Evidence that contradicts Speyside's theory	Inconclusive evidence

FIGURE 7.6 Chart to help summarise the field evidence for and against Speyside's hypothesis. List your observations in the appropriate column to help decide on your final recommendation to the next meeting at the Geological Society.

warming and the disappearance of all ice around the globe, but it would also need a lot more from somewhere. Another means of leading to an apparent sea-level rise is to actually subside the landmass. Stretching and thinning the Earth's continental crust can result in the upper part undergoing brittle failure. This can drop 'blocks' of former terrestrial areas below sea-level as part of a fault-bounded *basin*. This process of *rifting* can certainly happen quickly in geological time and cause vertical movement on the scale Speyside is suggesting (for instance, the East African Rift Valley). Not surprisingly, the fastest relative sea-level rise that you are likely to see preserved in the rock record will be a combination of eustatic change and subsidence.

At the bridge sit down to contemplate the pros and cons of Speyside's theory. Use Figure 7.6 to combine the sedimentological evidence about drift gathered in step one with the observations made on the 'erosion platform' above Garbh Allt and from the summit above Cnoc Breac.

- If Arran was drowned in the way that Speyside suggests, then what kind of sediments and fauna would you expect to see associated with this 'transgression'? (These should lie between the bedrock and the final cover of 'beach deposits', or drift.)
- The last volcano to erupt in this area of Scotland was 58 million years ago. What implication does this have on any notion Speyside might have of rifting being responsible at least in part for an apparent sea-level change?
- If Arran had subsided to give the sea-level change then why are you not swimming in the sea right now?
- What is your recommendation to the Geological Society about Speyside's hypothesis?
- What theory will you offer the meeting to explain the field evidence in Glen Rosa?

As you walk back to the *Start Point*, there is one last geomorphological feature worth pointing out. Approaching locality 1 (Figure 7.1) you should notice that Glen Rosa begins to open out into a wide, flat plain. There is a good view across it by the time you rejoin the road at the gate [NS 0015 3759]. This exercise has concentrated on geomorphological features caused by *either* ice or coastal erosion. The flat plain seen before you was in fact caused by both. During the Pleistocene this area would have been a sea loch similar in appearance to Loch Ranza today (on the northern tip of the island). The weight of ice loading the crust was so great during glaciation, that when it melted and this

FIGURE 7.7 There are two good raised beaches seen on Arran and the main road is built on the lowest and most prominent of these. The former sea cliffs can be seen to the left of the photo and the amount of rebound between these and where the sea-level is now can be estimated at about 7 m. Look out for old sea caves, stacks and arches as you travel around the island on this road.

weight was released, the land literally 'bounced' back up. This is called *isostatic rebound* and has caused an apparent sea-level fall (by raising the land-surface) of about 40 m over the last ~18 000 years, causing the loch of Glen Rosa to drain back to Brodick Bay. The rebound was incremental, so a series of separate levels, or *raised beaches* can be traced. The road around the island is built on one old beach and the former cliff-line can quite clearly be seen, often showing caves, stacks and arches cut by the waves (Figure 7.7). Speyside should have examined these raised beaches to see how different they are from the landforms high up in Glen Rosa that he was attributing to the same processes.

Help!

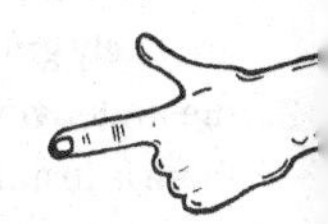

1 Figure 7.8. *(right)* Typical *drift* is an unconsolidated jumble of lithic fragments and smaller clasts in a clay matrix. Often the lithologies represented are foreign to the surrounding bedrock and can have a provenance hundreds of miles away.

2 Figure 7.9. *(overleaf)* The beach at Blackwaterfoot today demonstrates many of the features of a clastic shoreline.

The poorly sorted boulders, pebbles and quartz sand are continually pounded by the waves and tide increasing their maturity, sorting and roundness.

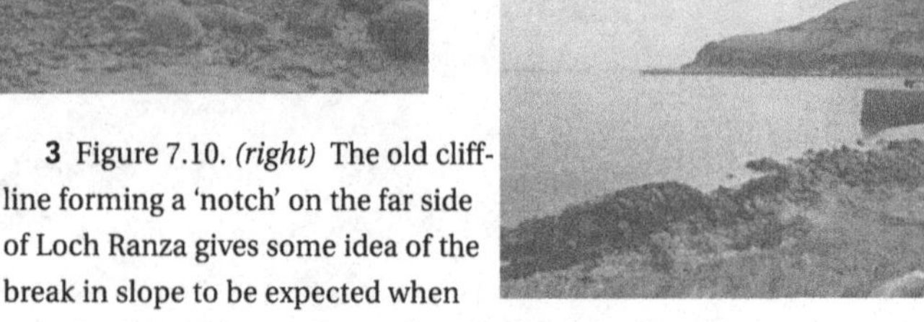

3 Figure 7.10. *(right)* The old cliff-line forming a 'notch' on the far side of Loch Ranza gives some idea of the break in slope to be expected when a sea-level is stationary for any length of time before falling.

(a)

4 Figure 7.11. At times even today it is not difficult to imagine ice covering much of Arran. In areas where ice build-up occurs it tends to take on one of two forms. (a) If there is sufficient snowfall and accumulation, then a blanket of ice will spread and cover the land. If this ice-sheet moves then it will gradually grind down and level-off the underlying bedrock. (b) Where there is insufficient snow, or further downhill, the ice tends to collect into 'tongues' or glaciers, confined within a 'channel' or valley. The erosive action of the ice everywhere in contact with the sides of the valley tends to carve a 'U' shape in profile.

(b)

Incidentally, Speyside's theory was wholeheartedly rejected at the Geological Society's emergency meeting and his integrity as a scientist was in tatters. He went on to become a Cabinet Minister with a seat on the Board of Directors at several major Industrial companies. Sir Stanley Speyside was knighted in 1997 and lives in Barking.

Exercise 7: field notes exact location	General location: Date:	data

Exercise 7: field notes exact location	General location: Date:	data

Exercise 7: field notes exact location	General location: Date:	data

Exercise 7: field notes exact location	General location: Date:	data

8 The curious case of 'Catacol cairn' * * *

(Catacol Bay, north-west coast)

PRELIMINARIES

Task To determine the geological origin of an offshore shingle bank known as 'Catacol cairn' which is formed of loose rocks not found *in situ* anywhere on the north-west coast.

Logistics *Start Point:* Lay-by 150 m north of Catacol Bay Hotel [NR 9109 4984]

The small village of Catacol lies in a broad bay on the north-west coast of Arran. The 'cairn' is on the foreshore at the northern end of the village near the Catacol Bay Hotel. Parking for a limited number of cars can be found in a lay-by on the roadside overlooking the cairn about 150 m north of the Hotel [NR 9109 4984]. From here there is immediate access to the exercise area and it is a good vantage point to assess the state of the tide before commencing work.

NB Tide time-tables must be consulted before attempting this exercise as the cairn is cut-off and completely submerged as the tide comes in. So only attempt this exercise when the tide is going out. This should leave plenty of time to finish work on the cairn and return to the roadside and complete the synthesis of field evidence. Waiting until the tide has turned and is coming in again before starting the exercise is too late.

Length

2 hr 30 min: Immediate access to the cairn, 2.5 hr to complete the exercise, Finish at *Start Point.*

Field skills required [1] [2] [3] [4] [6] [7] [8] [12] [13] [15] [16] [17] [21] [22] [25]

Background information

FIGURE 8.1 View north to Catacol cairn at low tide from the lay-by *Start Point*. The whole cairn is completely submerged at high tide. Note *in situ* outcrops of bedrock in the foreground which can be examined and compared with the boulders that compose the cairn.

'Catacol cairn' [NR 9100 4990 to NR 9115 5016] is a low bank or shoal of boulders and shingle that trends south-west to north-east across the foreshore of northern Catacol at low tide (Figure 8.1). Its presence on the beach might easily go unnoticed except for the fact that the boulders of which it is composed are neither the same lithology or the same age as the rock which outcrops along the shoreline on this side of Arran, or even across Kilbrannan Sound on the Kintyre Peninsula. Where these boulders have come from and how they came to form this bank is still problematic even today.

The cairn was first described by the Survey geologist, William Gunn (1837–1902) in the course of his work on Arran during the years 1892 to 1901. He was responsible for mapping almost the entire island single-handed, but died before the final memoir was published. Gunn's work was added to by Tyrrell in a subsequent edition, but essentially the BGS memoir still published today is that of Gunn's (a better testament to the geologist than his meagre obituary in the *Geological Magazine* of 1902). He had his own pet theory on the origin of the cairn and since that time others have been added. The principle ones that are relevant to this exercise are those of C. R. Cowie in 1905 and Murray Macgregor (author of the original and excellent excursion guide to Arran in 1965).

This exercise involves field skills introduced in Exercises 1, 2, 4 and 5 but it also involves some ideas presented in Exercises 6 and 7 regarding interpretation of sedimentary environments and glacial processes, respectively. It requires a limited amount of reconaissance mapping to begin with and then concentrates on age-dating the cairn from its faunal assemblage. This is related to Arran's stratigraphic column and the geological time scale. By considering three different theories which attempt to explain the existence of the cairn such diverse topics as longshore drift, glacial transport and 'Hutton's unconformity' are investigated.

THE FIELD EXERCISE

Read through all the instructions thoroughly before beginning work. Annotate maps and figures as required and write your answers to the questions as field notes on the pages provided at the end of the exericse. Remember, there's a 'Help!' section at the end in case you get stuck.

Steps needed to achieve the set task

1 Identify the bedrock along this stretch of coast.
2 Identify the different rock types that compose the cairn.
3 Assess the distribution of lithologies in the cairn.
4 Date the cairn using fossil and environmental evidence.
5 Locate possible source outcrops for the cairn boulders.
6 Synthesis: a hypothesis to explain the origin of the cairn.

Identify the bedrock along this stretch of coast

This first step of the exercise can be conducted whilst waiting for the tide to fall sufficiently to allow access to the cairn. The intriguing composition of the cairn cannot be put fully into context until the *in situ* outcrops along this coastline are identified to serve as a reference point. Since outcrops surround the cairn on all sides these need to be checked and identified to make sure that none of the lithologies on the cairn has come from on-shore exposures. Step from the lay-by *Start Point* onto the foreshore where outcrops occur directly seaward of it (Figure 8.2). Anywhere from here south to the jetty at [NR 9100 4962] can be examined.

- What lithological category do these rocks belong to? [2] [3] [4] [5] [7] [16]
- What is the lithological name for this rock? Use Figure 8.3 to draw a representative sketch of this lithology. [4] [5] [7] [16]
- What might the parent rock(s) have been? [16]

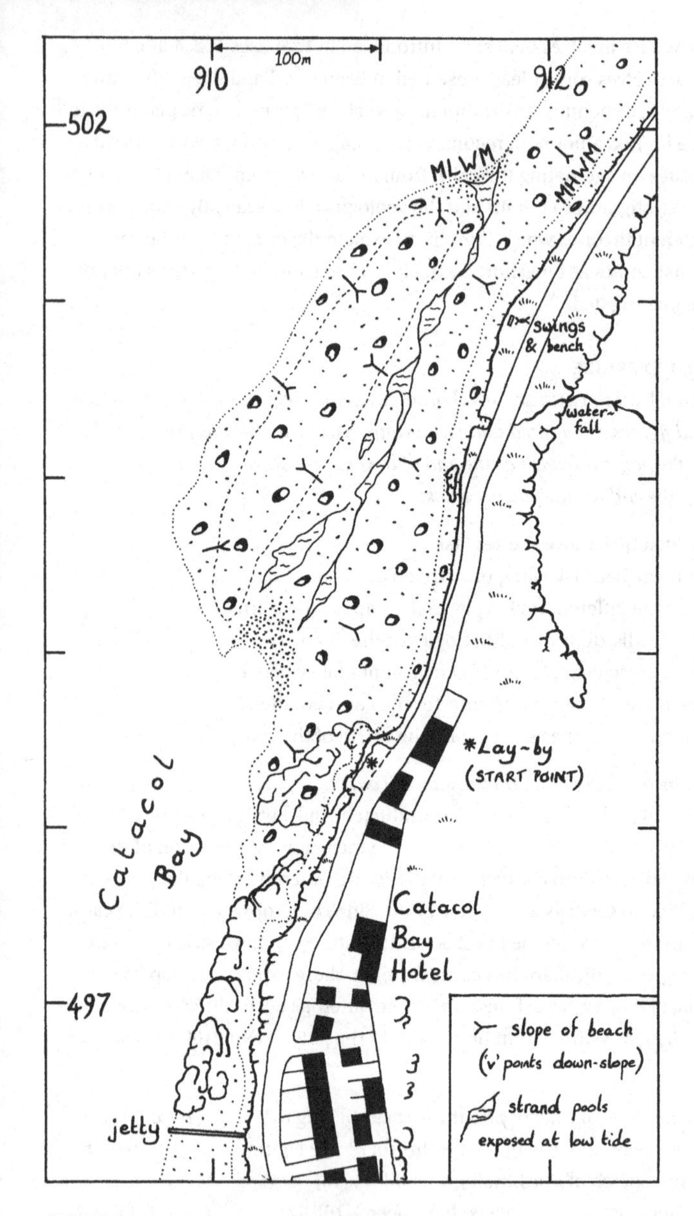

FIGURE 8.2 Exercise base map of Catacol cairn.

- A coarse-grained metasediment is said to have a 'psammitic' texture whilst the finer-grained equivalent is termed 'pelitic'. Are there any such banded variations in grain size/texture apparent in these rocks?
- What cleavage(s) do the rocks have? [15]
- Is there any folding and if so, how can the folds be described? [17]
- From the descriptions above, find these rocks on Figure 8.4. What is the age range of these rocks (in Ma)? (e.g. from 170 to 58 Ma)

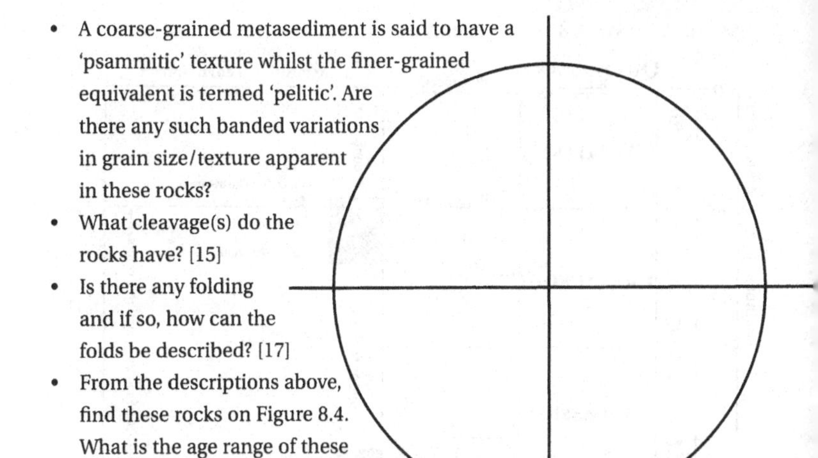

FIGURE 8.3 Imagine this circle printed on the rock, and draw in the mineral clasts that you see with a hand lens, making sure to show their physical properties such as size, roundness and metamorphic textures such as foliation. Then estimate the percentage of each mineral present within the circle. For instance, quartz = 60%, plagioclase feldspars = 40%. NB The total must equal 100%!

Pick a colour for this rock type, draw the outcrop shapes onto Figure 8.2 and shade them in [21]. Then walk north–north-east along the foreshore close to the road and stop to identify any other exposures, for instance there is one at [NR 9114 5000]. Repeat the same questioning procedure as above, colour the outcrop on the map and then cross the road to look at the cliffs near the bench and swings [NR 9122 5004], identify and mark these on too. This should provide convincing evidence that the bedrock in this vicinity is all the same lithological unit and the cairn can now be compared with this.

Identify the different rock types that compose the cairn

This step can be attempted when the tide has turned and the cairn is gradually appearing. Identifying the major lithologies that compose the cairn can be done by walking in a line at ~300°, perpendicular to the strike of the ridge (Figure 8.5) [1]. Start at the swings and follow this bearing off the grass, onto the foreshore and stop at the strand line (mark your position on Figure 8.2). Examine the boulder lithologies here (these are any lumps of loose rock ≥5 cm in size).

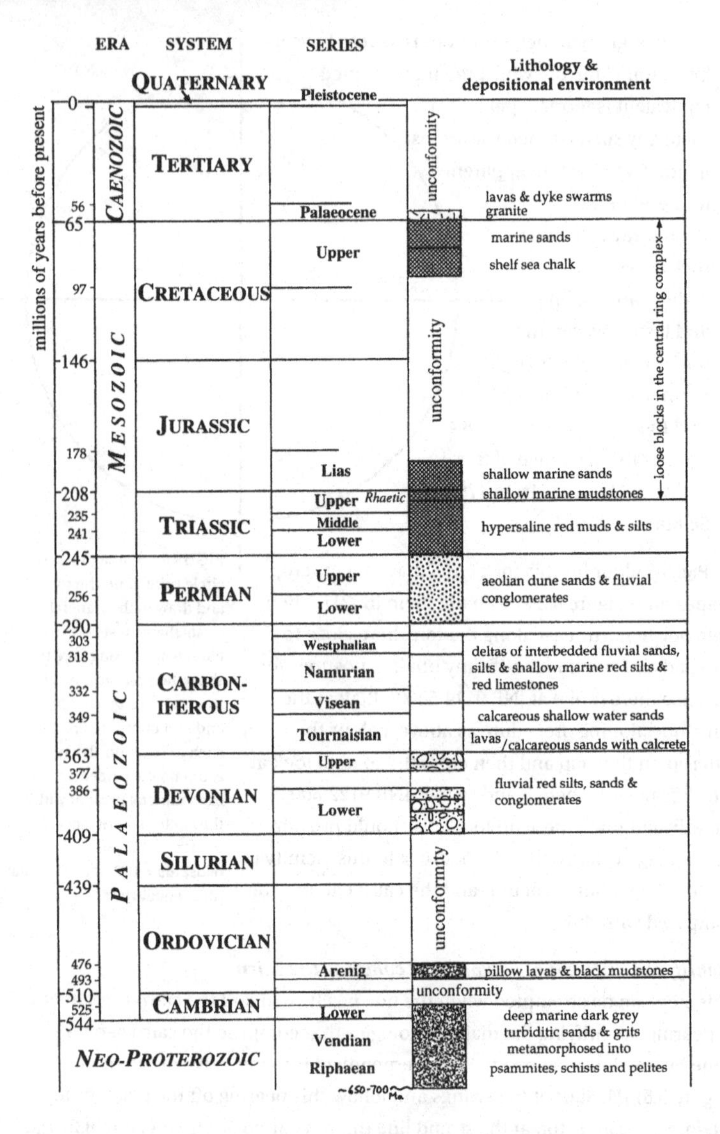

FIGURE 8.4 Simplified stratigraphic column for Arran. Using the description of lithologies and associated depositional environments, it should be possible to place boulders from the cairn in their correct stratigraphic position. Once this has been done, read off the appropriate age for the lithology.

FIGURE 8.5 Standing on the cairn ridge looking east towards the cliffs and swings at [NR 9122 5004]. The photo is taken along the line of transect perpendicular to the strike of the cairn. Note that there are no exposures present, just large boulders which show a change in lithology across the transect.

- What different lithological categories do the rock boulders belong to? [2] [3]
- What minerals can be identified in each different lithology of boulder? [2] [4]
- What is the name for each boulder lithology? [2] [3] [4] [5] [7] [16]
- In a 1 m square of ground near where you are standing, what would be the ratio of each boulder lithology to one another? (For instance, fine sandstone boulders 30%: 70% granite boulders. Make sure that they all add up to 100%).
- If any of the boulders are sedimentary in origin, do any of them contain fossils and if so, what is their name and to what group do they belong? [2] [8]

Once boulders at the strandline have been studied, pass through the narrow channel that separates the cairn from the shore onto the leeward side of the cairn ridge (e.g. [NR 9113 5012]). Again, mark your position accurately on Figure 8.2 and then examine the boulders here using the same questioning procedure outlined above.

- Is the ratio of different boulder lithologies to one another, or the lithologies themselves, any different from your description at the strand line, and if so, how?

Finally, to complete the transect, walk over the ridge to the seaward side. It is important to examine loose boulders exposed by the tide as far out as *safely* possible (at about [NR 9109 5014]). It may take a few mintues to 'get your eye in' and begin to recognise a lithology that was either not present, or was extremely rare along the transect so far. Follow along the edge of the water southwards down the cairn if after a while there is nothing obvious. This lithology should be easily recognisable because it is so completely different in almost all physical properties from the rocks composing the cliffs. Find, describe and identify a few examples of these *fossiliferous* boulders, again using the questioning procedure above. Remember to mark the position of your final boulder descriptions onto Figure 8.2. Once this is completed, your field observations from the three separate sites should demonstrate a change in boulder lithology, at least on the seaward side of the cairn.

Assess the distribution of lithologies in the cairn

It will help with the geological interpretation of the cairn's origin if the spatial distribution of the different boulder lithologies is known. The way to do this is to make a simple survey of the cairn and map the approximate zones where each of these principally occur. The bedrock outcrops on the foreshore are already marked on the map in colour. So pick another colour to symbolise each *dominant* boulder lithology, or lithologies, that can be recognised. For reference, mark these on as a key in the top left-hand corner of the map. Then walk the length of the cairn, first along the seaward side, then on the top of it and finally cover the leeward side and the channel area. Shade the main lithological zones on the map.

- Where are boulders matching the bedrock most common?
- Where do most fossiliferous boulders occur?

Date the cairn using fossil and environmental evidence

The structure and lithological variation within the cairn have now been documented, but what remains to be done is to place this into context with geological time and Arran's stratigraphy. This can be achieved using a combination of bio- and litho-stratigraphy. In other words, dating the rocks based on their fossil content and comparing depositional environments through time. So take a few minutes to scour the edge of the sea along the cairn and find as many fossils as possible. Study them with a hand lens and make sketches of each one that you find rather than collect the specimens themselves (use [8] to help draw and identify them).

- What is the name of each fossil found and to which group do they belong? [2] [8]
- Obviously the boulders have been moved, but are the fossils within them in 'life position' with respect to the matrix/cement? [2] [8]

- If the fossils were transported before deposition in the soft sediment, was it over a long or a relatively short distance, and what is the evidence for this? [2] [6] [8]
- Would the organisms originally have lived in freshwater or marine conditions? [8]
- Would the water have been deep or shallow? [8]

Consider these interpretations in conjunction with Figure 8.4. The rocks exposed on Arran can be conveniently split into six major litho- and chrono-stratigraphic units, as shown by the shading on Figure 8.4. You should already have identified the stratigraphic position, and consequently the age, of the bedrock at Catacol on this figure.

- In what Geological System(s) could the lithology which forms the fossiliferous boulders of the cairn have originally been deposited?
- Which 'Series' have the appropriate depositional environment for the fossil assemblage in the boulders? (Shade these in on Figure 8.4 to help with the next question.)
- What is the age range of the fossiliferous boulders in Ma?
- How much time separates the deposition of the bedrock from that of the fossiliferous boulders (Ma)?

Locate possible source outcrops for the cairn boulders

Having dated the fossiliferous cairn boulders, walk back to the roadside to avoid the possibility of being cut-off by the tide. The exercise task stated that the (fossiliferous) boulders of the cairn were of an age and lithology unlike any others on the north-west coast of Arran. By now it should be clear that this is certainly true at least as far as Catacol Bay is concerned. To look further afield for the outcrop from which they are derived requires a map not just of this stretch of coast, but of the whole island and the rock that outcrops under the sea as well. It is important to realise that 'geology' does not stop at the waters' edge and that as far as this exercise is concerned the sea is a modern-day inconvenience obscuring our view of what is beneath. Luckily, Figure 8.6 is just such a map.

- What areas on Figure 8.6 could the cairn boulders have come from? (Shade these in the same colour you used on Figure 8.4 to designate their age.)

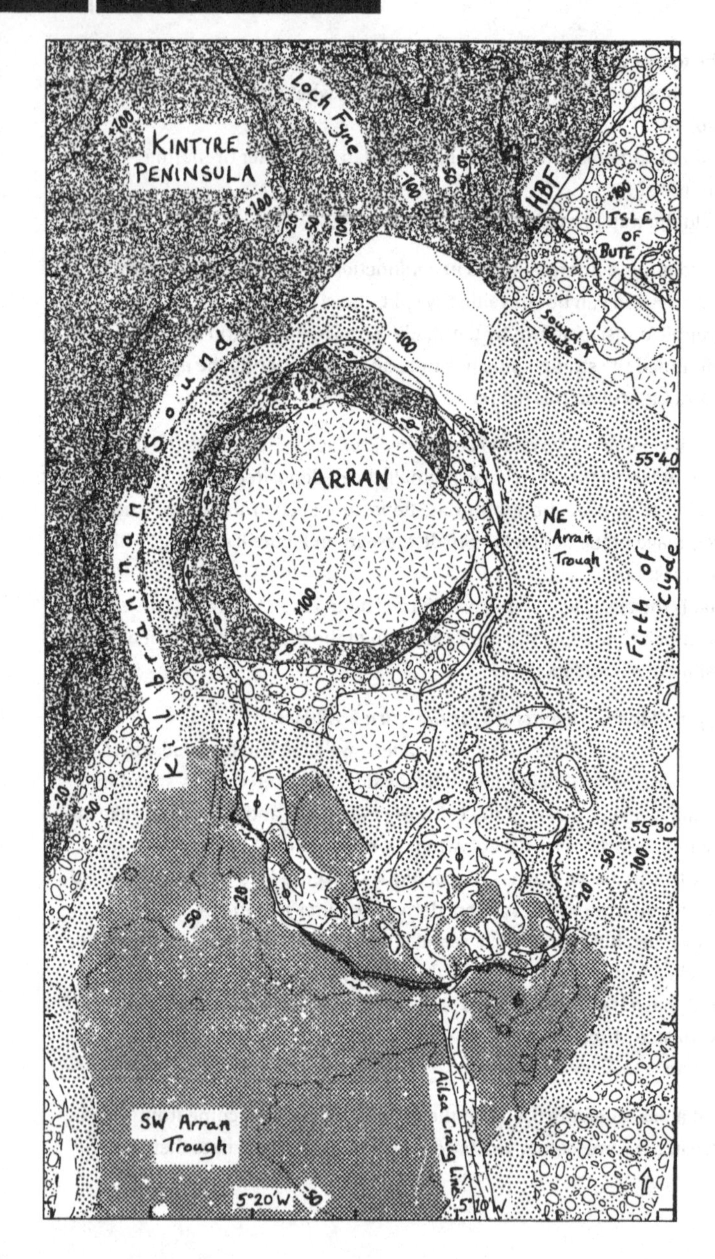
Loch Fyne
KINTYRE PENINSULA
+100
+100
-20
-50
-100
-100
-50
-10
HBF
+100
ISLE OF BUTE
Sound of Bute
-100
Kilbrannan Sound
ARRAN
+100
55°40
NE Arran Trough
Firth of Clyde
-20
-50
55°30
-20
-50
100
-20
-50
SW Arran Trough
5°20'W
Ailsa Craig Line
5°10'W

Synthesis: a hypothesis to explain the origin of the cairn

Figure 8.6 demonstrates the problem of explaining where the fossiliferous cairn boulders have come from. Consider all the field evidence that you have collected so far, and then look through these three hypotheses that have been proposed as a solution in the past.

1. *Gunn's*: A narrow outcrop striking parallel to the coast lies just off-shore and storms bring boulders from it onto the ridge. If this hypothesis were correct then:

- How could an outcrop of that age lie stratigraphically on top of the observed *in situ* bedrock along this coast; what has happened to the thickness of rocks (and time) that should separate the two?

2. *Cowie's*: An ice-sheet moving south from Loch Fyne down Kilbrannan sound in the Pleistocene brought the boulders from further north (off the map). If this hypothesis were correct then:

- Why are the boulders not a random mixture of lithologies, perhaps even including some that are not seen on Arran? Also, why is there a zonation to the boulders across the cairn east to west?

3. *Macgregor's*: Longshore drift has brought the boulders round the northern tip of Arran from the on-shore and off-shore outcrops along the north-east coast. If this hypothesis were correct then:

- Why is there only one predominant lithology on the seaward side of the cairn; the tide should should have washed round boulders from a variety of units including Permian aeolian red sandstones?
- Which theory best fits your field evidence to explain the origin of Catacol cairn?

FIGURE 8.6 *(opposite)* A simplified geological map of Arran and surrounding areas with *projected* submarine outcrop (i.e. there is a good chance that it could be wrong!) and bathymetric contours. In a map such as this the present coastline almost becomes immaterial, but it can be seen that the east of the map corresponds to the Firth of Clyde, Loch Fyne is to the north and on the west is the Kintyre Peninsula. The south-west Arran trough opens out into deep-water towards Northern Ireland. The map shading corresponds to the six stratigraphical units on Figure 8.4. Also shown are the directions of ice movement during the Pleistocene (circles with a line through them), major tidal currents (hollow, outline arrows) and longshore drift directions around Arran's coast (wavy arrows, or wavy line with tick if no longshore drift is present). The reason for these being on the map will become clear as a hypothesis to explain the cairn is pieced together. This map is partly based on information in McLean, A.C. and Deegan, C.E. (eds.) 1978. The solid geology of the Clyde Sheet (55° N 6° W). *Rep. Inst. Geol. Sci.*, No. 78/9.

Incidentally, there is another shingle bank, or cairn, on the North Newton Shore at Loch Ranza [NR 9310 5135] that is reported to be composed of the same fossiliferous lithology as Catacol cairn. When the author visited this at mid-tide all the boulders not immediately derived from the local bedrock were Permian New Red Sandstones ...

Help!

1 *Gunn's theory*: Figure 8.7 *(right)*. How could a much younger outcrop lie stratigraphically on top of the bedrock you have described along this coast? This is an outcrop known as 'Hutton's unconformity' which lies on the North Newton shore at [NR 9340 5190]. Here, the lower unit has a cleavage which dips steeply to the left, but is overlain by *basal* Carboniferous carbonate-rich sandstones which dip to the right. The author's hand rests on the apparent contact of the two, but the cleavage can be seen above this level so the sharp junction may actually be faulted. NB An unconformable surface can have topography, so the thicknesses of younger units which are deposited on top can vary ([9]). For more information on Hutton and the importance of this outcrop, see the background information to Exercise 9.

2 *Cowie's theory*: Typical *drift* is an unconsolidated jumble of lithic fragments and smaller clasts in a clay matrix. Often the lithologies represented are foreign to the surrounding bedrock and can have a provenance hundreds of miles away. See Exercise 7.

3 *Macgregor's theory*: Longshore drift in Catacol Bay suggests a north-east movement of shingle along this stretch of coastline.

Exercise 8: field notes exact location	General location: Date:	data

Exercise 8: field notes exact location	General location: Date:	data

Exercise 8: field notes exact location	General location: Date:	data

Exercise 8: field notes exact location	General location: Date:	data

9 Dr Hutton's dilemma * * * *

(North Glen Sannox, north-east Arran)

PRELIMINARIES

Task To observe and assess the field evidence used by James Hutton to support his *Theory of the Earth* in 1795. Is his epitaph; *'the founder of modern geology'* justified?

Logistics *Start Point:* North Glen Sannox car park [NR 9932 4680]

North Glen Sannox is situated just inland to the north-west of Sannox Bay on the north-east coast of Arran. It can be reached via the main road through the valley connecting the hamlet of Sannox to Lochranza on the north coast. The nearest bus stop is in Sannox from which it is about a 15 minute walk north along the road into North Glen Sannox. Alternatively, it might be worth asking the bus driver to be let off at North Sannox bridge. The exercise area covers about 3 km^2 in North Glen Sannox and about another 1 km^2 over the watershed into Glen Sannox to the south. North Sannox Burn drains eastwards through the glen and bisects the exercise area. In the centre is North Sannox bridge and the *Start Point.* This is the only dry crossing point for the main burn in the map area. Two sets of deer fences are currently present in North Glen Sannox and these should *only be crossed at the stiles.* To the south of the main burn the valley is open, but to the north-east is a farm. There will occasionally be sheep in this corner of the exercise area so it is important to be considerate and *keep clear of any livestock,* especially if it is the lambing season. This exercise can be done in most weather conditions, but if it is particularly bad then it is advisable to keep down off the higher areas where visibility can easily be lost in cloud; there are few landmarks to navigate from once out of the valley bottom.

Length 1 to 1.5 days: 20 min walk to begin first step, Up to half a day to complete the first step. Remainder of time to complete the exercise.

Field skills required [1] [2] [3] [4] [5] [6] [7] [9] [10] {12] [13] [14] [15] [16] [17] [20] [21] [22] [23] [24] [25]

Background information

James Hutton (Figure 9.1) has been credited by many as the 'founder of modern geology' for the ground-breaking ideas set out in his now classic book, *Theory of the Earth*, published in 1795. The field work that finally convinced him of the validity of his theory was carried out on Arran. This exercise returns to some of those areas visited by Hutton in 1787 to reassess the field evidence.

FIGURE 9.1 James Hutton (1726–1797) in the field, the year he visited Arran. This is from a contemporary cartoon by John Kay. Notice the rather fashionable field clothing and that a geological hammer was already recognised as a crucial piece of equipment. It is amazing to think that Hutton would have looked like this when he came to North Glen Sannox in 1787.

In the late eighteenth century, the widely accepted theory for the origin and evolution of the Earth was that of Abraham Werner (Figure 9.2). According to Werner, all rocks had been deposited or crystallised in a few thousand years from an ancient, receding ocean that had originally covered the entire globe. They had been formed in a continuous and systematic order, with granite as the oldest rock underlying all others (Figure 9.3). This theory soon became christened *neptunism* (after the Roman god of the sea) and introduced the concept of *stratigraphy* to geology. However, Hutton, an ex-farmer and self-styled Edinburgh philosopher, did not subscribe to this Wernerian view of the planet.

Hutton had begun work in the 1740s on his own ideas of how the rocks of the Earth had evolved. But it took until the spring of 1785 before he finally made his ideas public by presenting a 'Theory of the Earth' to the newly formed Royal Society of Edinburgh. His 'Theory' proposed two staggering and totally radical ideas for the time. Firstly, Hutton suggested that land and mountains were continually

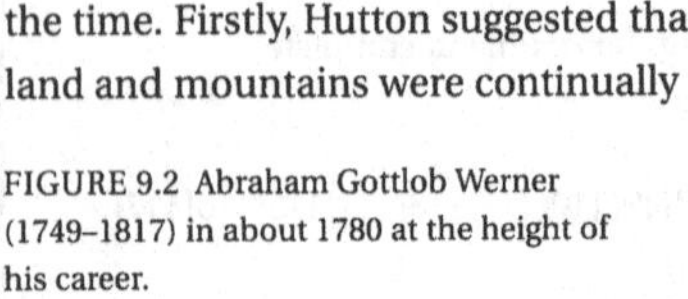

FIGURE 9.2 Abraham Gottlob Werner (1749–1817) in about 1780 at the height of his career.

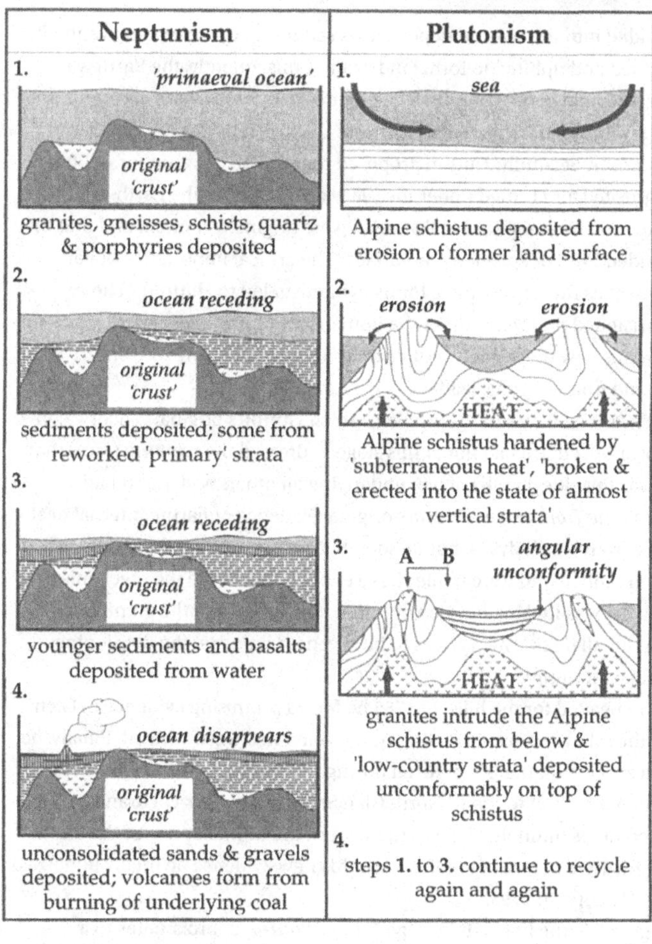

FIGURE 9.3 The two opposing theories of *Neptunism* and *Plutonism* side by side for comparison. Hutton travelled extensively around Scotland and divided the rocks that he found into four broad stratigraphic groups: *alpine schistus, low-country strata, granite* and *unconsolidated soil*. Using these divisions, Hutton thought that the Earth had experienced *at least* three cycles of marine deposition–uplift–erosion. The first of these was represented by the now badly **heated** and **deformed** *alpine schistus* (which formed the mountains of Scotland). Evidence of the second were the *low-country strata* (**undeformed** sediments from the Midland Valley in which he often found fossils). The third cycle was currently in progress with the erosion of soil today. Wherever he travelled Hutton found evidence of granites apparently intruding the rocks so he used this as the mechanism of uplift needed to complete a rock cycle. The erosion that took place at the end of each cycle, followed by renewed deposition, produced what Hutton called an *angular unconformity* in the rock record (see [9]).

being eroded into the sea and deposited as sediment, which would eventually be deformed and uplifted to form land again. Consequently, the Earth was dynamic and land could continually be renewed in these cycles *ad infinitum*. The Earth was therefore incredibly old, and this concept was the discovery of *geological time*. Secondly, Hutton proposed that the mechanism for uplifting rocks deposited and buried under the sea was heat within the Earth. He was convinced that he had found rocks which were originally molten, and that they had intruded from deep below the surface. The critical importance of Earth's internal heat as the driving force for its dynamism led to Hutton's 'Theory' being christened *plutonism* (after the Roman god of the 'underworld'). Not surprisingly, this new theory drew criticism from those adhering to Werner's philosophy and from religious leaders.

There was one very serious problem which Hutton's original 'Theory' completely overlooked. He had not at this stage addressed the *neptunist* view that *granite* was the oldest rock known, underlying all others and that it had *crystallised-out from water*. Hutton's original evidence of Earth's internal heat had come from basalt dykes, but he soon became convinced that granites had once been molten and were bodies large enough to provide the mechanism of uplift that he was looking for. He knew that he had to solve the granite issue once and for all if his 'Theory' was to be accepted amongst the developing geological community.

Hutton headed for the hills. In 1785 he found promising evidence in Glen Tilt and the following year in Galloway, but remained unconvinced. Finally, he achieved success on the Isle of Arran during the summer of 1787. The field evidence that Hutton found in North Glen Sannox and near Lochranza confirmed in his mind that he was right about the immense age of the Earth and the intrusive nature of granite. With this new evidence he finally published a revised 'Theory' in 1795.

In Figure 9.3, the line A–B in stage 3. of *plutonism* approximates to a cross-section west to east down North Sannox Burn, so Hutton must have realised that this glen contained all the elements of his 'Theory' in one place. This exercise sets about investigating the original field evidence that Hutton found and used before passing judgement on whether or not he deserves the reputation afforded him by historians of geology. To do this means constructing a map and cross-section of North Glen Sannox and Glen Sannox in order to produce a geological interpretation that can be compared with Hutton's. Was his judgement clouded by the need to find evidence in support of his 'Theory'?

THE FIELD EXERCISE

Read through all the instructions thoroughly before beginning work. Annotate maps and figures as required and write your answers to the questions as field notes on the pages provided at the end of the exericse. Remember, there's a 'Help!' section at the end in case you get stuck.

Steps needed to achieve the set task

1 Identify Hutton's units.
2 Reassess the stratigraphic succession.
3 Construct a geological map of the area.
4 Draw and interpret a cross-section.
5 Synthesis.

Identify Hutton's units

To begin the exercise, Hutton's units of *granite, alpine schistus* and *low-country rocks* must be found and identified in North Glen Sannox. Look again at Figure 9.3 if you are unsure of the physical properties that Hutton attributed to each map unit. The best way of identifying Hutton's lithologies is to begin in the west near the head of the valley and walk a transect down North Sannox Burn (Figure 9.4). *This should take no more than half a day to complete.*

Start in the burn close to A on Figure 9.4 and walk east, stopping to describe each new rock type along the side of the burn until you reach B at the right hand side of the map. Make sure to note down the physical properties of each lithology so that you would recognise it again when mapping up on the hill-sides where the exposure is not as good. Use the questioning procedure below to identify the lithologies of each new unit in the sequence. There are a few more than the three simple ones that Hutton recognises and it is important to make the distinction between them.

- To which lithological category do the rocks belong? [2] [3]
- What different types of mineral/lithology can be identified as crystals/clasts and what is the approximate size range of each? [2] [4] [5] [6] [16]
- What are the physical properties of the matrix or groundmass in each rock unit? [2] [5] [6]
- If the rock is sedimentary in origin, is the rock matrix supported or clast supported? [2] [7]
- What is the proportion of the whole rock that each mineral or lithic fragment represents in view under a hand lens? [2]
- What is the name of the lithology? [2] [5] [7] [16]
- Do the rocks have any cleavage(s)? [15]

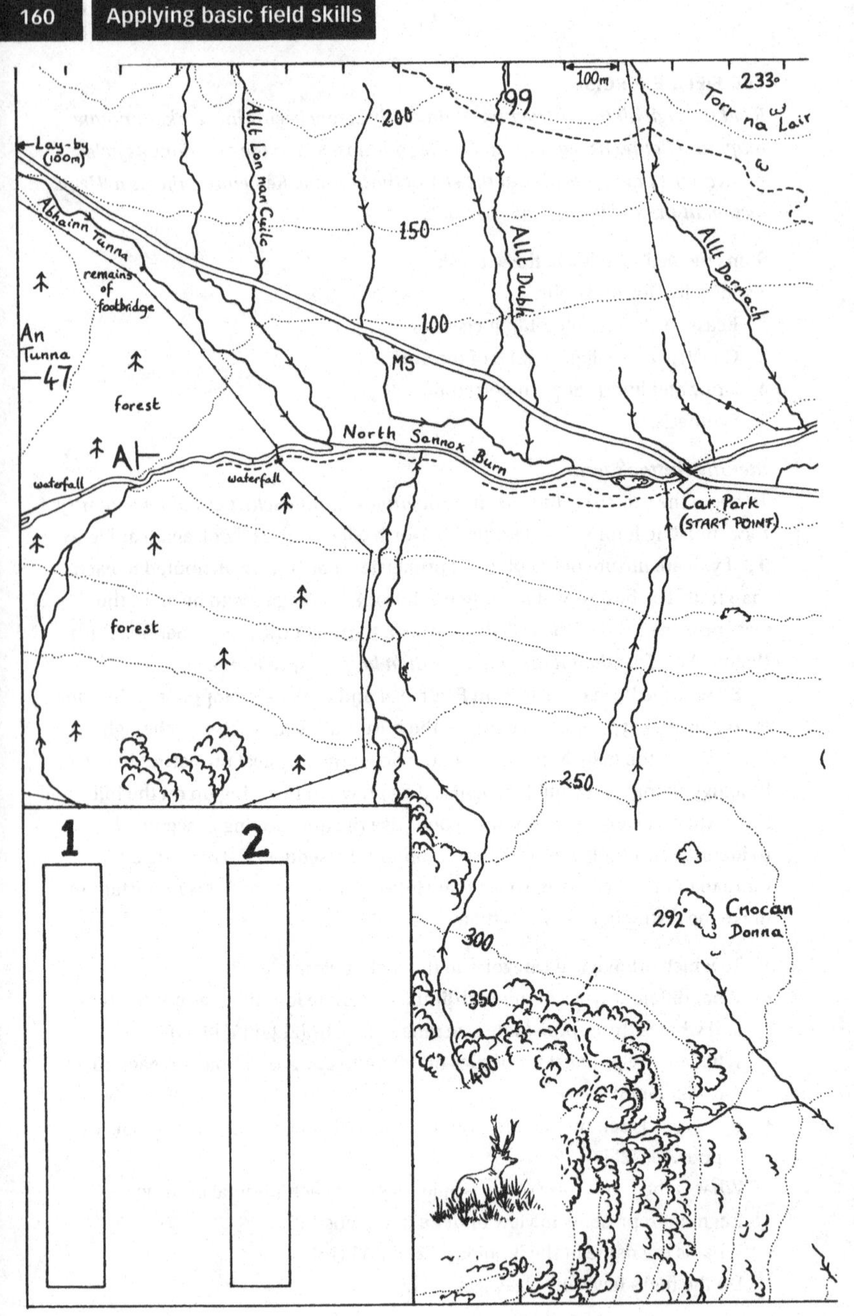
100m
233°
Torr na Lair
99
200
Lay-by (100m)
Allt Lòn nan Cuile
Abhainn Tunna
150
Allt Dubh
Allt Dornach
remains of footbridge
100
An Tunna
47
MS
forest
North Sannox Burn
A
waterfall
waterfall
Car Park (START POINT)
forest
250
1
2
292°
Cnocan Donna
300
350
400
550

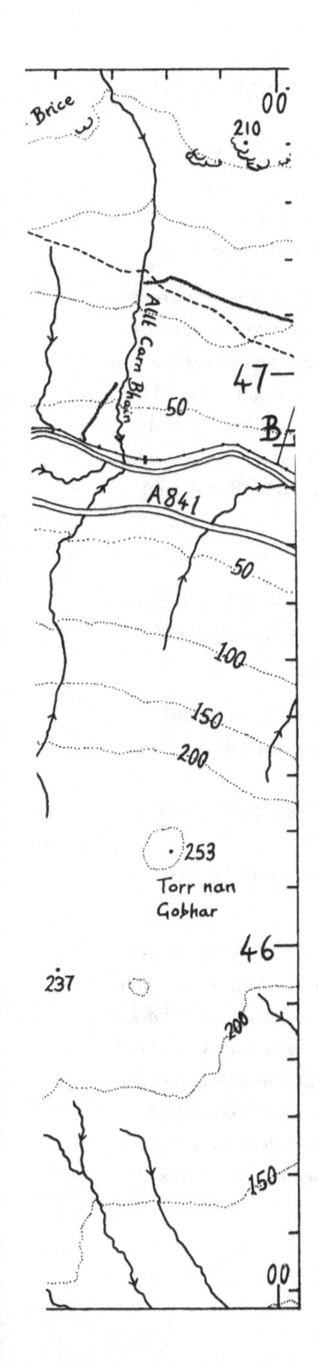

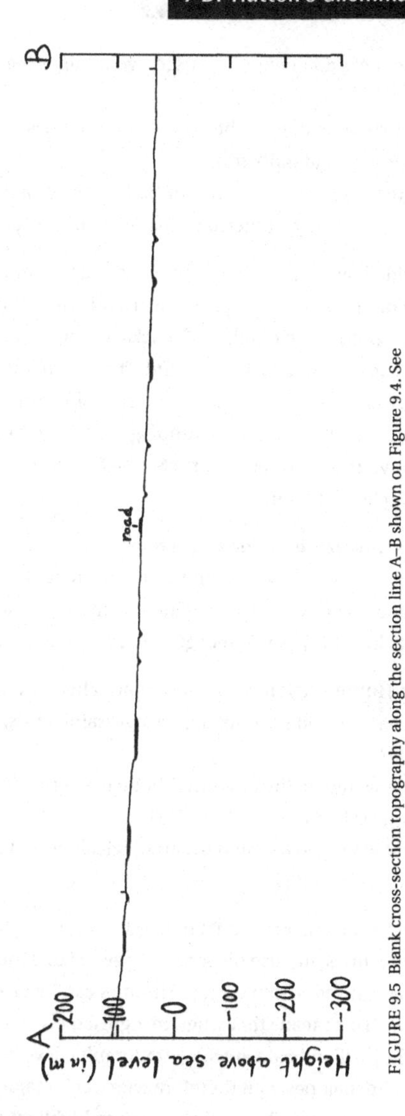

FIGURE 9.5 Blank cross-section topography along the section line A–B shown on Figure 9.4. See [24] on how to construct the section (only steps (viii) onwards will need to be done).

FIGURE 9.4 Exercise base map of North Glen Sannox and Glen Sannox. The position of cross-section A–B is shown and the two blank stratigraphic logs are labelled 1 and 2.

- If the rocks are metamorphosed, what might the parent rock have been? [16]
- In which direction do the rocks young? [9] (Use remnant sedimentary structures to identify this.)
- Are there any obvious faults or folds present in the rock? [17]
- Are there any hydrothermal 'quartz' veins present in the rocks? [17]

Decide how many different lithologies are in the map area and choose a pencil colour to represent each one. Divide the column of Log 1 (Figure 9.4) into the appropriate number of blocks to represent them. Starting at the bottom of Log 1, colour each block and write the name you are going to call the lithology next to it (use the left hand side of Figure (ii) as a guide if you are unsure). You *must* mark each lithology onto Log 1 in the order that you found them in your survey down North Sannox Burn. The reason for this should become clear later on.

Reassess the stratigraphic succession

Once the transect down North Sannox Burn has been made and the rock units described, the next step is to relate this modern interpretation to that of Hutton's in 1787. He only recognised three map units in this valley.

- Was Hutton right to keep his 'alpine schistus' rock undifferentiated?
- If it can be split into smaller, recognisable units, how many of them are there?
- What would be the modern lithological terms for these units in Hutton's 'alpine schistus'? [2] [4] [6] [7] [16]
- What would be the modern lithological term(s) for Hutton's 'low-country strata'? [2] [4] [6] [7]

Bracket together the units in Log 1 that belong to each of Hutton's three rock categories and use his terminology to label them. The new units that have been recognised within each of Hutton's can be used as stratigraphic *marker horizons*. This means that when recognised elsewhere in the valley at an outcrop, they will indicate at what stratigraphic level within the 'alpine schistus' the locality belongs, and therefore whether a major geological boundary should be close by. They will also help to highlight any large-scale structural deformation within this unit, a feature that Hutton used to characterise the 'alpine schistus'.

Construct a geological map of the area

In North Glen Sannox, Hutton thought he found field evidence for his sedimentary 'rock cycles' (see Figure 9.3) and molten intrusion of these by granite at a later date. Strictly speaking, in Hutton's view there should be an angular unconformity between the 'alpine schistus' and 'low-country strata' and granite should have intruded both of them (also thereby creating unconformable contacts with each). The only way to assess whether he was right or not is to produce a three-dimensional reconstruction of the geology in this glen (see Exercise 5). The next two steps lead you through this process by first constructing a plan view of the units on the ground, and then by making a cross-section interpretation through the valley.

Map-out the boundaries between the new units you recognised in the last step, across the whole exercise area [21] [22] [23]. Where you start doing this is up to you. As might be expected in an area such as North Glen Sannox and Glen Sannox, the best, or most complete exposure is to be found in stream beds and in particular along North Sannox Burn itself. *Allow yourself anything from half a day to a day to complete this step.*

During mapping, be sure to mark on dip and strike of geological boundaries ([10] and [21]) wherever possible as these are needed to construct a cross-section [24]. The map should be accompanied by field notes, sketches and structural information ([12], [13], [15], [17], and [20]). Critical localities are present at the three farthest corners of the map so it is vital to cover the whole area shown and not just concentrate efforts along the burn. Turn to the *Help!* section to find sketches of the most important localities that should be found in the field. Answer the questions below when you think you have completed the map.

- Can the contact of the 'granite' and the 'alpine schistus' be seen in the map area, and if so what is the nature of the contact?
- If this has not been observed, can the relationship of the two units to each other be inferred from the outcrop patterns? (This is where mapping the *marker horizons* within the 'alpine schistus' is particularly helpful.)
- Can the contact of the 'alpine schistus' and 'low-country rocks' be seen in the exercise area, and if so what is the nature of the contact?
- If this has not been observed, can the relationship of the two units to each other be inferred? What are their general dips and dip directions?
- What are the problems in measuring the angles of dip/dip directions in the 'low-country strata'?
- So where are Hutton's major unconformities on the map? (Draw their trace at the surface in a distinctive colour; see [9])

Draw and interpret a cross-section

The plan-view of a geological map must be combined with a vertical 'slice' through the rock to gain an overall idea of the three-dimensional structure in the exercise area. So, construct a scale cross-section along the line A–B on Figure 9.5. This should aim to bring out and clarify the main points that contribute to understanding the geological history of the valley, and in particular the nature of the boundaries between Hutton's units. The questions below might help with this interpretation.

- Which are the oldest sediments?
- Which are the youngest sediments?
- What are the youngest rocks in the area?
- Have any, or all, of the sediments been metamorphosed or deformed? (Which units show cleavage?)
- Which way do the rocks 'young', and is this all in a uniform direction or does the younging direction reverse? [9]
- If the younging direction reverses then how might this be explained? [25]
- How does the cross-section compare with the line A–B on Figure 9.3?

Look again at Log 1 on Figure 9.4. Log 2 next to it is still blank, so now fill this in with your revised stratigraphy of the map area, placing the oldest rocks at the bottom and working up to the youngest rocks at the top of the log.

Synthesis

Armed with your geological map and cross-section, return to considering *Neptunism* versus *Plutonism* with these final questions.

- Were *all* the map units in North Glen Sannox and Glen Sannox laid down gradually and without interruption over time as the Neptunist theory would maintain?
- What exactly was it that Hutton found at the 'granite'–'alpine schistus' contacts that questions the Neptunist view of granite and gives an idea of its relative age to the 'schistus'?
- Is it possible to find the relative age of the granite to the 'low-country rocks' in the map area as Hutton implies you can?
- Are there any eroded fragments of 'alpine schistus' contained within the 'low-country strata' and how might this affect the validity of Hutton's 'Theory'?

- What other piece of field evidence can still be used to infer the 'low-country strata' were formed in a younger 'cycle' of uplift and erosion?
- Based on your own observations in the map area, which aspects of Hutton's theory are supported by the field evidence?
- Based on your own observations in the map area, which aspects of Hutton's theory do you disagree with and why?
- Was Hutton a good field geologist?
- In your opinion does Hutton deserve his reputation for overturning the neptunist ideas with his 'Theory of the Earth'?

Two hundred years on, and based largely on research over the last fifty years, geologists would now envoke a quite different mechanism of uplift from that proposed by Hutton. His field observations were generally sound though, and interestingly enough Arran's northern granite did dome-up the surrounding sediments as it intruded. However, the large scale uplift of ocean floor onto continental landmasses is best explained these days by collision of tectonic plates. The deformation and uplift of Hutton's 'alpine schistus' actually occurred during the Ordovician and Silurian as the continental masses of England and Scotland collided, destroying an intervening ocean that once existed between the two. Slivers of this former ocean floor were thrust up and plastered onto the Scottish continent and evidence of this is seen in North Glen Sannox. Hutton's overlying 'low-country strata' are a mixture of both marine and non-marine sediments essentially eroded off the landmass produced during collision, and he was right to spot that they are relatively undeformed (tilted, yes, but metamorphosed, no). Granite, as he decided from the field evidence, was indeed the youngest rock in North Glen Sannox. Just to make sure this was right, Hutton hauled his colleague, John Clerk, all the way round the side of Goatfell tracing the 'granite'–'alpine schistus' contact back to Brodick. From Hutton's own account of his trip to Arran in 1787, he says that he set out alone on horseback one day from Brodick to Lochranza (see why in Figure 8.7), stopping in North Glen Sannox on the way. It is an intriguing possibility that he discovered both geological time and Earth's internal heat within a few hours of each other. Not a bad day in the field.

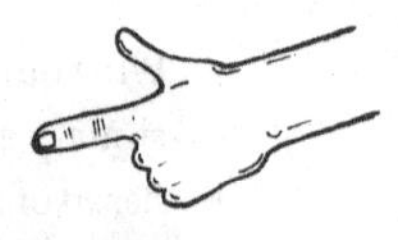

Help!

1 Figure 9.6 *(right)*. The 'granite'–'alpine schistus' contact in North Sannox Burn. The author is looking at small structures which Hutton recognised and used as evidence for the relative age of the granite. The photograph faces south and was taken in summer when the burn was low.

2 Figure 9.7 *(left)*. Field sketch of one of the units that Hutton failed to recognise within the 'alpine schistus' rock on the northern bank of North Sannox Burn, looking west. The altered, elongate basalt (spilite) 'pods' display glassy rims, vesicles and convex surfaces which suggest that they are pillow lavas (see [13]) and give a younging direction for this unit. *If you can find this locality in the field, annotate the sketch with your geological observations.*

3 Figure 9.8 *(right)*. Field sketch of sub-rounded altered basalt (spilite) clasts in a gritty matrix; north bank of North Sannox Burn looking north. What could be the origin of this lithology? *If you can find this locality in the field, annotate the sketch with your geological observations.*

4 Figure 9.9 *(right)*. Field sketch of an irregular 'granite'–'alpine schistus' contact in a stream west of Cnocan Donna; drawn facing east. *If you can find this locality in the field, annotate the sketch with your geological observations.*

5 Figure 9.10 *(left)*. Field sketch of repeating coarse, fine and graded units in Hutton's 'alpine schistus' demonstrating the original sedimentary bedding and the younging direction, northern flank of Torr nan Gobhar. *If you can find this locality in the field, annotate the sketch with your geological observations.*

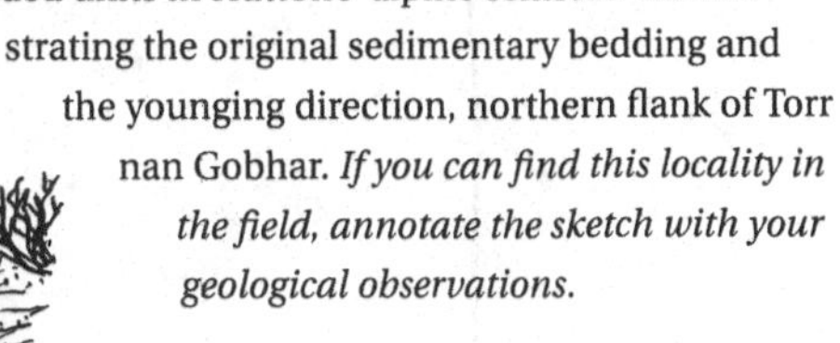

6 Figure 9.11 *(right)*. Plan view field sketch of sub-angular altered basalt (spilite) clasts in a dark grey/black 'shaly' matrix with strong planar fabric; summit of Cnocan Donna. This suggests that the 'feature' in Figure 9.8 is mappable across the valley-side. *If you can find this locality in the field, annotate the sketch with your geological observations.*

Exercise 9: field notes exact location	General location: Date:	data

Exercise 9: field notes exact location	General location: Date:	data

Exercise 9: field notes exact location	General location: Date:	data

Exercise 9: field notes exact location	General location: Date:	data

Exercise 9: field notes exact location	General location: Date:	data

10 Pluto's revenge ****
(Drumadoon, west coast)

PRELIMINARIES

Task At Drumadoon there is a complex of cross-cutting dykes and sills. Your task is to establish the order in which they were intruded.

Logistics *Start Point:* Kinloch Hotel car park, Blackwaterfoot [NR 8958 2813]

The exercise area covers about 1.5 km^2 in grid squares NR 88 28 and NR 88 29. Either catch a bus or drive to the car park outside the Kinloch Hotel in Blackwaterfoot [NR 8958 2813] and then walk west into the exercise area along the coast road to Shiskine golf course. On reaching the club-house [NR 8915 2845] cut down onto the beach and continue along towards the southern 'feeder' dyke [NR 8850 2869] to start the exercise. After only a few hundred metres two basic dykes strike across the beach and these can be found in the bottom right-hand corner of the exercise map. *All field work can and should be completed without straying onto Shiskine golf course.*

NB The Doon is bounded on its western side by vertical columnar-jointed cliffs. The composition of the Doon can be examined at its base via a path that runs below the columns. However, the cliffs are unstable and the nesting sea-gulls frequently kick rocks down onto the path below. Thus, if this path is to be taken it is important to: (a) wear a hard hat, and (b) not hammer anything in the vicinity. Exposures on top of the Doon are best reached via a path at its northern extremity [NR 8866 2948].

Length 1 to 1.5 days: 20 min walk to begin first step, Remainder of time to complete the exercise, 20 min walk back to *Start Point*

Field skills required [1] [2] [3] [4] [5] [10] [12] [13] [14] [15] [17] [21] [22] [23] [24] [25]

Background information

This exercise demonstrates the techniques needed for successful field mapping and subsequent interpretation of emplacement history in a complex series of acid and basic intrusions around the Drumadoon area. The intrusions took place at depth before being 'unroofed' during the late Tertiary, and are now exposed amongst near-horizontal Triassic interbedded hypersaline muds, silts and fine sandstones. The Drumadoon complex perhaps displays some of the best examples of 'magma mixing' to be seen in this country. The intrusions form intricate associations of phenocrysts, 'xenocrysts', 'xenoliths' and groundmasses in a variety of remarkable combinations. The best explanation for their formation lies in the mixture of processes that can occur in a magma chamber and during emplacement (Figure 10.1).

Exercise 3 and [5] both address the classification of igneous rocks in the field mainly as a simple function of crystal size and mineral suites. In most cases this 'rule of thumb' is sufficient for a field description. Unfortunately, this is not so at Drumadoon because intrusive melts from acidic to basic compositions have mixed at various stages of differentiation from the magma chamber to their subsequent emplacement. This has produced many *porphyritic* rocks, which usually take their name from the large crystals that can be identified in them. For instance, a common igneous rock on Arran and one seen in Exercise 3 is a 'quartz–feldspar porphyry'. This method of classification is not adequate enough to describe the intrusions at Drumadoon. The only way to classify the Doon rocks in such a way that will demonstrate their problematic associations and thus aid interpretation, is to give each rock a name based on both the phenocrysts/xenocrysts and the groundmass together. So, instead of a *quartz–feldspar porphyry*, it should be called a *quartz–feldspar felsite*. This may seem pedantic before field work commences but it should quickly become clear that there are many rocks at Drumadoon which have large quartz and feldspar crystals in them but completely opposite groundmass compositions. Figure 10.2 gives some lithological names that may become useful in the Drumadoon area.

This exercise is slightly different from the others that involve geological mapping as the steps here lead the user to seven key localities. Step 8, the synthesis, is then to make sense of these isolated pieces of the jigsaw puzzle and put them together into a 3-D reconstruction of the Drumadoon intrusions. It does not matter in which order steps 1 to 7 are tackled, but it is important that the whole area is mapped. The steps are simply critical localities that must not be missed.

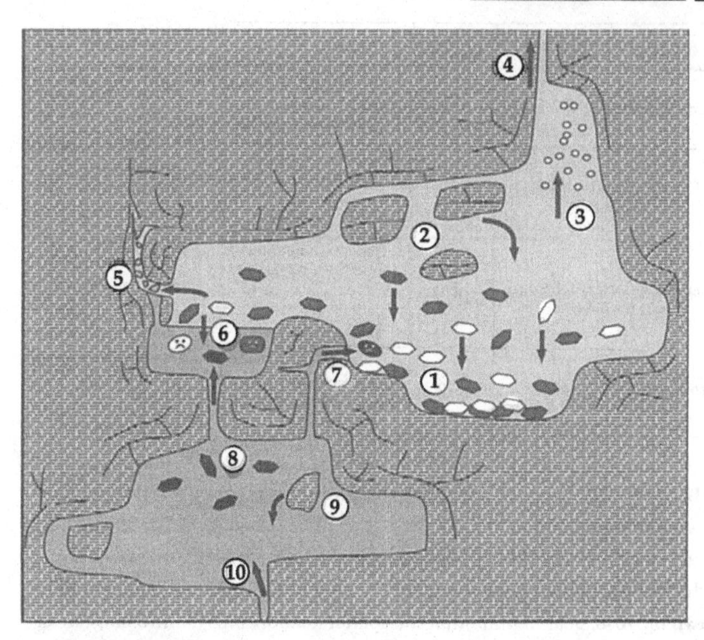

FIGURE 10.1 A stylised, 'ideal' magma chamber demonstrating some of the ways in which the composition of magma might be 'fractionated', and how melts or crystals of two distinct compositions could be mixed before, or during emplacement. It is a hypothetical situation which helps to explain how some of the features seen in the Drumadoon area may have formed. 1. The large magma chamber begins to cool as it moves towards the surface and crystals start to form. They may sink to the base of the chamber or crystallise around its edges. The remaining magma will have its composition changed by the loss of elements needed to form the crystals. 2. Blocks of country rock may be 'stoped' and fall into the chamber as it progresses upwards. These are termed 'xenoliths'. They may completely melt and be incorporated into the magma, so slightly changing its composition. 3. The magma will have a number of volatiles within it; given the chance these will escape out through the top of the chamber perhaps through fractures, changing the composition of the melt they leave behind. 4. As crystals are forming, magma may be allowed to escape upwards through cracks in the country rock, producing dykes. This will change the composition of the melt left behind. 5. An acidic melt escaping from a chamber via a fracture already occupied by a cooling, more basic melt could result in xenoliths of the basic material being caught up and intruded along with the acidic melt. 6. Crystals from one melt, maybe a silica-rich one, may sink into a more dense, basic melt below. The silica-rich crystals are 'unhappy' in a melt that is Si-deficient and will begin to dissolve (see [5] for which mineral associations might be expected in silica-rich and silica-poor igneous rocks). These 'foreign' crystals are termed 'xenocrysts'. 7. The reverse situation of 6.; ferromagnesian minerals will reabsorb into a more acidic melt if the temperature is still high enough. 8. The original magma chamber may be partly refilled from below by a magma of a different composition, perhaps a more basic one with basic phenocrysts. The two may stay separated as immiscible liquids in the chamber. 9. Xenoliths of country rock may be stoped into a more basic magma in a similar situation to 2. above. 10. The system may be refilled from below by a melt of composition unlike either of the two shown to produce even more lithological variety.

Full lithological name	*Map symbol*	*Description*
Quartz alkali feldspar felsite	**QFF**	Quartz & *alkali* feldspars in a groundmass of devitrified pitchstone
Quartz alkali feldspar andesite	**QFA**	Similar to **QFF** except groundmass more intermediate in composition
Quartz plagioclase feldspar basalt (or dolerite)	**QFB**	Quartz (sometimes rounded) & *plagioclase* feldspars in a fine to medium crystalline, basic groundmass
rounded alkali feldspar basalt (or dolerite)	**rFB**	*Alkali* feldspar reabsorbed into basic melt before cooling halted process

FIGURE 10.2 Chart showing some of the ways in which to describe both the phenocryst and groundmass minerals in mixed melts. There are more compositions than just the four shown here, but these can be named using a similar method. The reasons for choosing the title of this exercise should now be clear!

Use [14], [21], [22], [23] and [25] to help you construct a geological map of the area in Figure 10.3, and [24] for cross-sections when you have finished. It is particularly important to notice when intrusions cross-cut or chill against each other as this will allow a relative chronology for their emplacement to be drawn up. Also, bear in mind that the spatial relationships of intrusions are as important as their composition. Always measure the dip and dip direction of contacts and the trend of dykes wherever possible and mark them in accurately on the map (see [10] and [14]). Lastly, it is perhaps worth mentioning that you should pay special attention to the complex relationships of intrusions around the 5th Tee (Figure 10.3).

THE FIELD EXERCISE

Read through all the instructions thoroughly before beginning work. Annotate maps and figures as required and write your answers to the questions as field notes on the pages provided at the end of the exericse. Remember, there's a 'Help!' section at the end in case you get stuck.

Mapped locations needed to achieve the set task

(NB Steps 1 to 7 can be done in any order)

1 Southern 'feeder' dyke.
2 Drumadoon Point.
3 The 5th Tee.
4 The Doon stack.
5 The base of the Doon.
6 Fallen columns.

7 The top of the Doon.
8 Synthesis.

Southern 'feeder' dyke [NR 8850 2869]
This is a good opportunity to become accustomed to a common lithology in the map area. Look at the overall shape and form of the dyke, then have a closer look at it through a hand lens. The sea has eroded many fresh surfaces on it, so it should be fairly easy to see and identify the phenocrysts. Use Figure 10.4 to help describe this lithology.

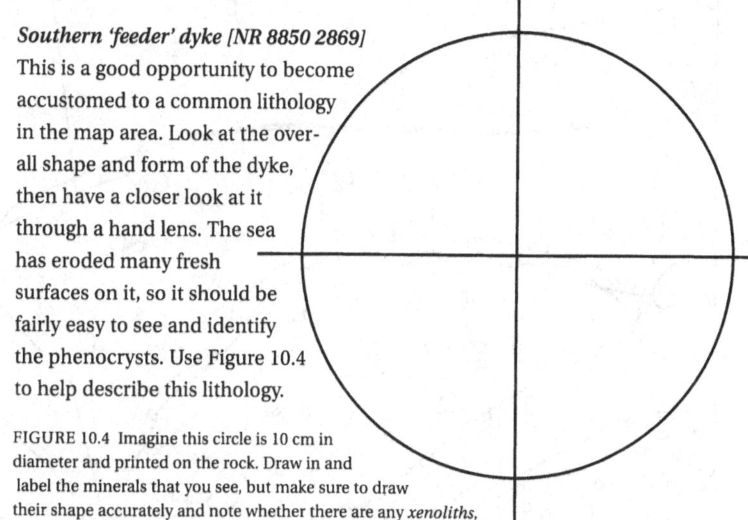

FIGURE 10.4 Imagine this circle is 10 cm in diameter and printed on the rock. Draw in and label the minerals that you see, but make sure to draw their shape accurately and note whether there are any *xenoliths, xenocrysts, twins, overgrowths* or even *xenocrysts* present (see [5]). Then estimate the percentage of each mineral present within the circle. For instance, quartz = 60%, plagioclase feldspars = 40%. NB The total must equal 100%!

- What is the name of this rock and why? [2] [3] [4] [5]
- What is the attitude of this dyke; i.e. what is the dip and dip direction of its contacts with the country rock? [10]
- What is the trend of this intrusion; why might it be called a 'feeder' dyke?

Drumadoon Point [NR 8825 2879]
The sands are constantly shifting at this locality to reveal and cover the country rock. If it is exposed and the tide is low then it reveals a gully with an off-set between two inclined sills on either side (Figure 10.5). Whether the country rock is exposed or not, it is worth spending a few moments to examine the subtle difference in composition of the two sills and the nature of the contact between them.

- How does the composition differ between the two intrusive units? [2] [4] [5] Use Figure 10.6 to help describe their mineralogical differences.
- What is the dip and dip direction of the contact between the two sills? [10]
- In which order were the two sills intruded? (Use Figure 10.6 and [5] to help you decide.)

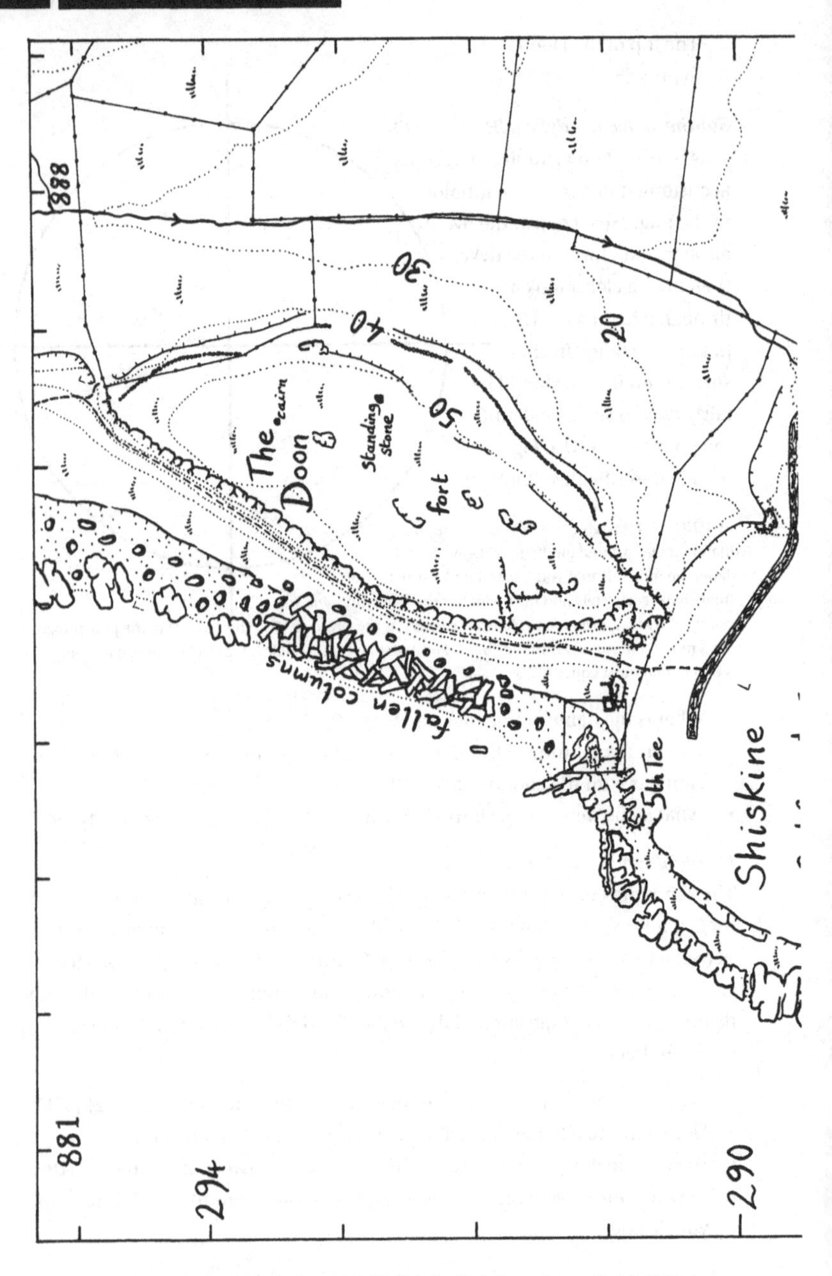
888
881
294
290
30
40
50
20
The Doon
cairn
Standing stone
fort
fallen columns
5th Tee
Shiskine

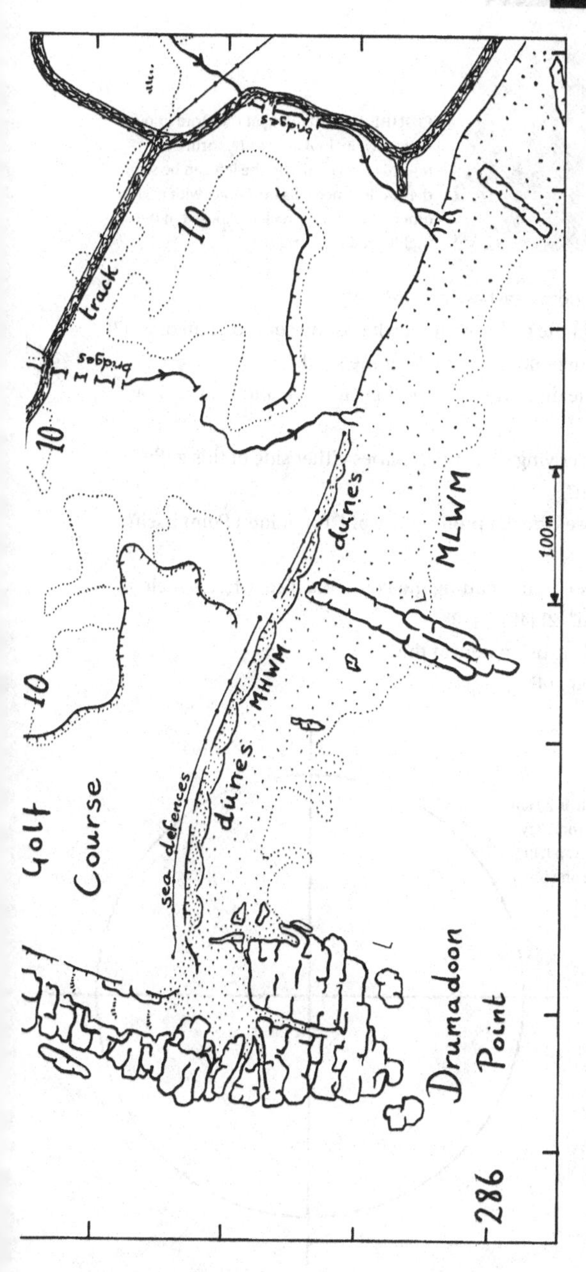

FIGURE 10.3 Exercise base map for the Drumadoon area. The box just north-east of the 5th Tee shows the approximate outline of the enlarged sketch map of Figure 10.7. NB There are no exposures on the golf course so there is no need to stray onto it.

FIGURE 10.5 Standing at the Drumadoon Point locality looking north–north-east towards the Doon. On the left can be seen the two inclined sills in contact with each other. The country rock is exposed in the middle and to the right.

- Why is there a gully trending east–west here? [14]
- Is there any exposure in the gully and if so, what is the composition of it? [2] [4] [5] (This will again depend on the state of the sand.)
- Does the gully offset the dipping contact either side of it and if so, by how much?
- The thickness of the overlying sill actually varies either side of this gully. How could this happen?
- What lithology composes the flat promontory of Drumadoon Point itself? [2] [4] [5]
- Are there any other dykes cross-cutting the Point and if so, what is their composition and trend? [2] [4] [5] [10]
- What was the order of intrusion here at the Drumadoon Point area? [14]

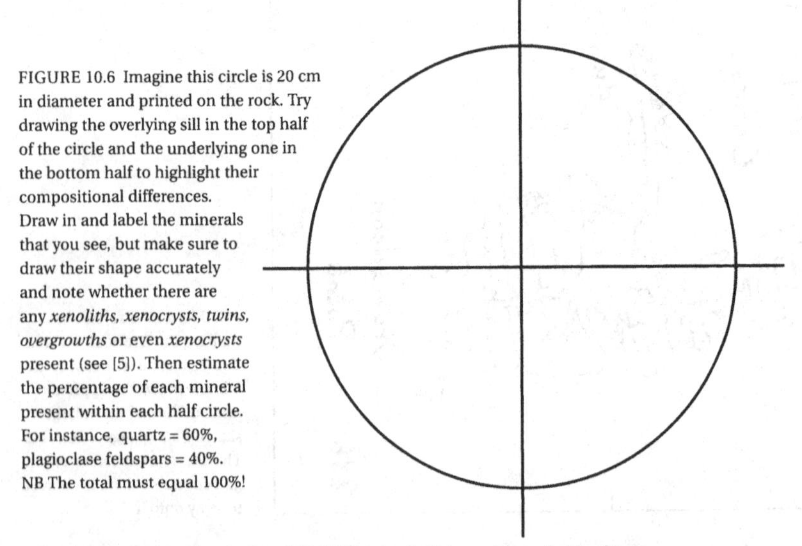

FIGURE 10.6 Imagine this circle is 20 cm in diameter and printed on the rock. Try drawing the overlying sill in the top half of the circle and the underlying one in the bottom half to highlight their compositional differences. Draw in and label the minerals that you see, but make sure to draw their shape accurately and note whether there are any *xenoliths, xenocrysts, twins, overgrowths* or even *xenocrysts* present (see [5]). Then estimate the percentage of each mineral present within each half circle. For instance, quartz = 60%, plagioclase feldspars = 40%. NB The total must equal 100%!

The 5th Tee [NR 8833 2907 to 8843 2913]
In many respects this small area is the key to understanding how the Doon relates to Drumadoon Point and so needs to be mapped in some detail. However, be prepared for it to take a while as there are several intrusive phases here and the lithologies are by no means clear. To help you work out what is going on, Figure 10.7 shows an enlarged area on the foreshore in front of the Doon. Switch to using this larger scale map when things begin to get a little crowded on Figure 10.3. You may decide that you need to draw a similar style sketch map for the western part of the 5th Tee area.

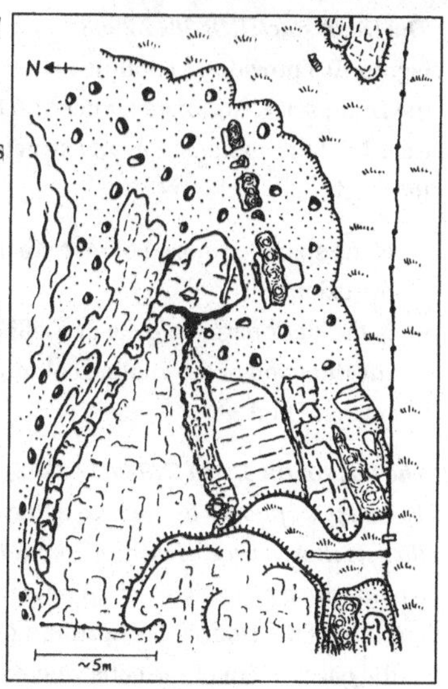

FIGURE 10.7 Enlarged base sketch map showing part of the 5th Tee area. The map shows the wire fence along the edge of the golf course and in the bottom right-hand corner, the gate which must be crossed to gain access to the Doon foreshore. This sketch map was drawn in the field and so the scale is variable. The principal jointing in the intrusions and country rock is also shown.

- What is the order of intrusive phases in the 5th Tee area? [2] [4] [5] [10] [14] (Use Figure 10.8 to sketch some of the mineralogical compositions and textures that you encounter here.)

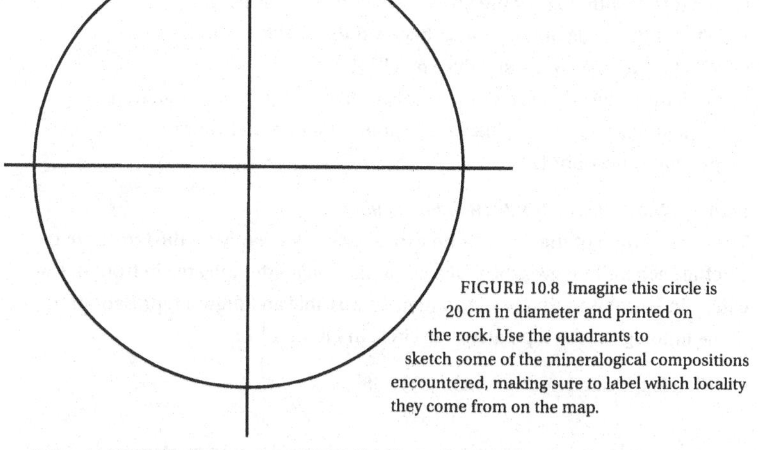

FIGURE 10.8 Imagine this circle is 20 cm in diameter and printed on the rock. Use the quadrants to sketch some of the mineralogical compositions encountered, making sure to label which locality they come from on the map.

The Doon stack [NR 8847 2908]

Figure 10.9 provides an outline field sketch of the Doon stack. Annotate it once you have examined the composition and structure of the stack.

- Is the stack composed of more than one intrusion?
- Is the stack part of a dyke or a sill? (Look at the direction of the cooling columns for this answer [14].)

FIGURE 10.9 Outline field sketch of the Doon Stack. Fill this in with the appropriate lithologies and be sure to mark on the direction of the cooling columns [14].

The base of the Doon [NR 8847 2915]

This step is optional as the base of the cliffs can be dangerous. If it is attempted, then as short a time as possible should be spent actually at the cliff-face. There is a path which can be picked up near the stack and runs along under the western face of the Doon to its northern tip. This allows examination of lithologies in the basal 1–2 m of the cliffs where they are in contact with the Triassic country rock below. So follow the path along to about [NR 8847 2915] and then scamble up the bank to the base of the cliffs. *Only attempt this if there are no nesting birds above and you have a hard hat.* The lowermost ~1 m of the cliff has a different composition than the main mass of columns. Concentrate on this lower region here and use Figure 10.10 to help describe it. The lithology of the main part of the Doon can be observed in the fallen columns on the foreshore during the next step.

- What is the lithology of the lower ~1 m of the Doon? [2] [3] [4] [5]
- What is the angle of the contact between this and the country rock?
- Was the Doon a composite dyke or sill? [14]
- Look up at the vertical columns that compose the cliffs. If these were formed during cooling, does this support or contradict your answer to the previous question? [14]

Fallen columns [NR 8841 2919 to 8849 2934]

The composition of the main Doon can be examined easily on the foreshore by carefully scrambling over the fallen columns which litter the area in front of the cliffs. Find a column that is as complete as possible and draw a representation of the lithology in the top half of the circle in Figure 10.10.

- What is the name for this rock? [2] [4] [5]

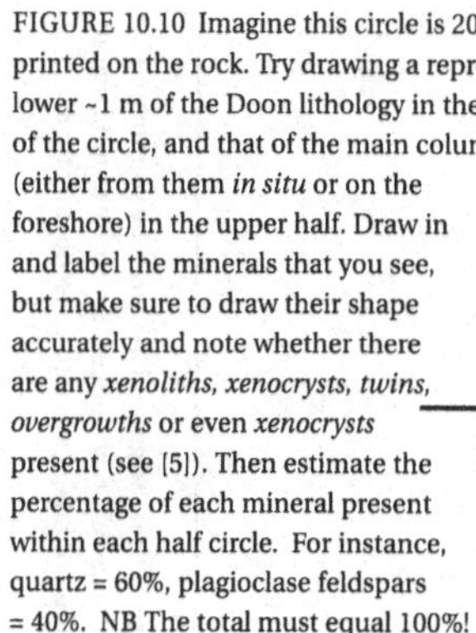

FIGURE 10.10 Imagine this circle is 20 cm in diameter and printed on the rock. Try drawing a representation of the lower ~1 m of the Doon lithology in the lower half of the circle, and that of the main columns (either from them *in situ* or on the foreshore) in the upper half. Draw in and label the minerals that you see, but make sure to draw their shape accurately and note whether there are any *xenoliths, xenocrysts, twins, overgrowths* or even *xenocrysts* present (see [5]). Then estimate the percentage of each mineral present within each half circle. For instance, quartz = 60%, plagioclase feldspars = 40%. NB The total must equal 100%!

- What are the xenoliths composed of? [2] [3] [5]
- How does the lithology of these central columns compare with that at the base of the Doon?
- Have you seen any evidence which allows the relative timing of emplacement between the fallen column lithology and the basal layer of the Doon to be established? (Look at the sketches you have drawn of the two rock types and refer again to Figure 10.1.)

The top of the Doon [NR 8850 2906 to 8865 2946]

Popular opinion has the Doon capped as well as underlain by a ~1 m thick sill of a composition unlike the middle of it. Survey the top of the Doon with this idea in mind. The southern half yields some interesting exposures.

- Is there any evidence of this top sill?
- Are there any dykes cross-cutting the Doon, and if so, what are they composed of? [2] [4] [5] [10] [14]

Synthesis

Complete the geological map by interpreting and then inking-in the course of intrusion boundaries. The map needs to be accompanied by cross-sections to demonstrate the sub-surface geology. This will allow you to interpret the three-dimensional structure of the Drumadoon dyke and sill complex. Selecting appropriate lines of section across the map area is difficult but it is suggested that two parallel sections are constructed east–west; one across the Doon itself, the other across the Drumadoon Point area. No blank cross-

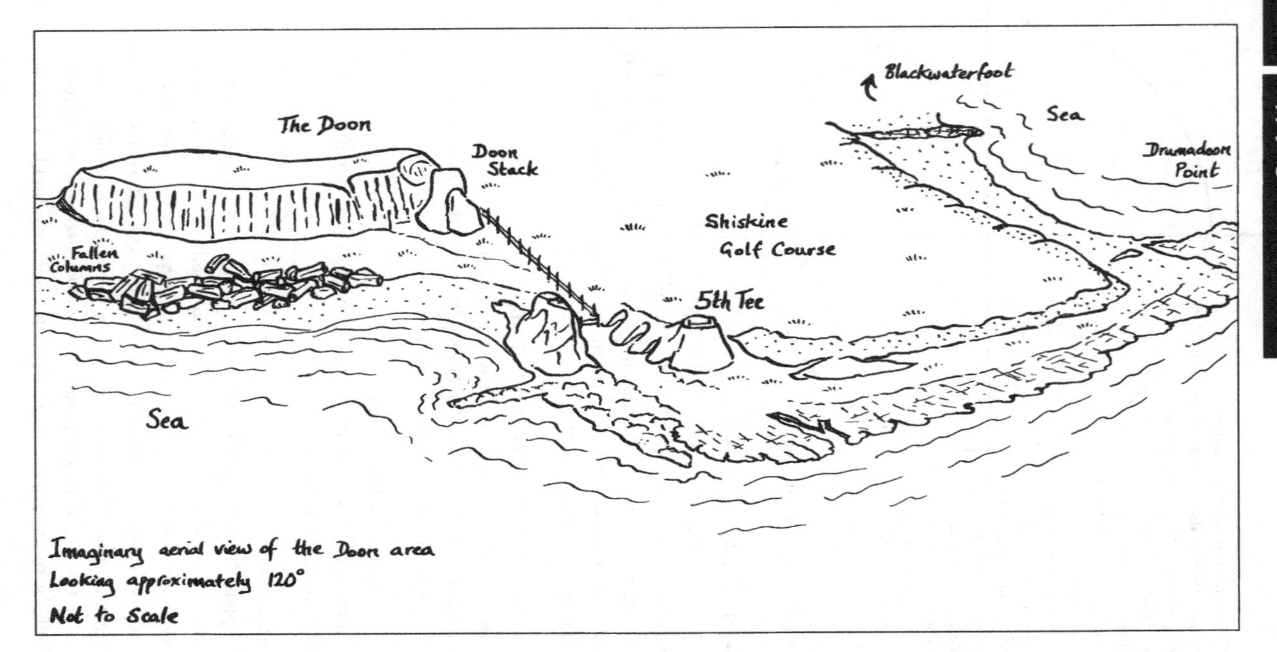

FIGURE 10.11 Bird's eye view of the Doon complex. Start by drawing the approximate boundaries of dykes and sills on the ground as shown by your map. Then, imagine that the country rock is transparent and try drawing on the 3-D picture of what the dyke and sill complex would have looked like above ground before it was eroded away. Use your cross-sections to help with this, and in doing so, try and think about which intrusions cut other ones and consequently their relative ages.

section templates are provided for this last exercise; drawing cross-sections from scratch is described in [24].

One of the main questions frequently asked about the intrusions in this area is whether the Doon and Drumadoon Point are part of the same intrusion which has been thrown by an east–west oblique-slip fault [17], or whether they were intruded at different levels in the first place. Look back over your field evidence, the map and the two cross-sections. If all seven localities have been visited then you should be able to offer an explanation. It is often difficult to visualise the 3-D structure of an area in the mind. To help with this, use Figure 10.11 to draw what you think it would have looked like before it was eroded down to present-day topography. Also, try answering these questions about the relative phases of emplacement, and the two together should help you solve the Drumadoon igneous complex.

- Which intrusions were dykes and which were sills?
- Which younger injections have chilled against an already cooled, or cooler, older intrusion?
- Which intrusions have been cut through (or displaced) by younger ones?
- Did Drumadoon Point have the same composition as the Doon?
- What did the orientation of cooling columns in the Doon stack show?
- What shape in cross-section do you think the Doon was originally?
- Were there any features indicating faulting near the 5th Tee, such as slickensiding, fault breccias or veining? [17]
- Was there more than one phase of acidic magma emplacement in the area?
- Was there more than one phase of basic magma emplacement in the area?
- Were there any igneous bodies of a hybrid composition and how might they have arisen?
- So what was the sequence of igneous intrusion in the Drumadoon area?

Help!

1 Figure 10.12 *(right)*. Contact in cross-section between the two inclined sills at Drumadoon Point. *If you can find this locality in the field, annotate the sketch with your geological observations.*

2 Figure 10.13a *(below)*. Close-up of the overlying sill at Drumadoon Point. Note the rounded shape of some crystals and that

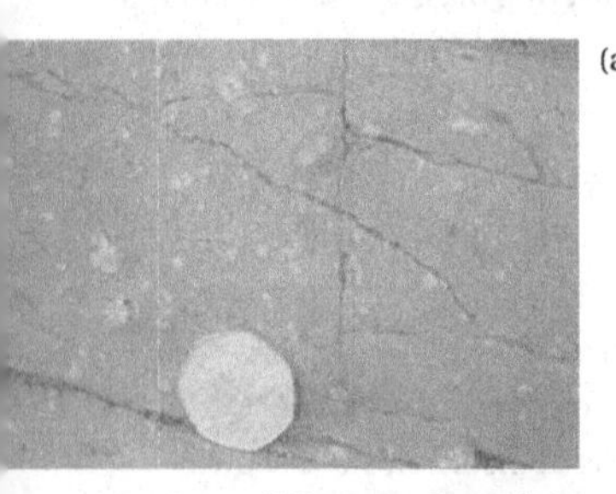

(a)

(b)

some have different coloured rims. These are plagioclase feldspar rims around xenocrysts (see [5]). Figure 10.13b *(right)*. Close-up of the underlying sill at Drumadoon Point. Note the euhedral shape of the crystals and that the groundmass is lighter than in the overlying lithology (see [5]).

3 Figure 10.14 *(right)*. Looking west from the 5th Tee. Gullies often indicate where dykes have been weathered and eroded out. So tracing the course of gullies across the foreshore is a useful mapping procedure.

4 Figure 10.15. View to the north down onto the foreshore from the 5th Tee on the golf course. (a) *(right)* Looking north–north-east, there is a complex set of dykes beneath the covering of boulders in the centre of this photo.

(a)

(b)

(b) *(left)* Looking north-east towards the enlarged sketch map area. The dyke complex passes through the gully between the low stack and the edge of the golf course. Try drawing the course of the dykes that you find directly onto these photos with coloured pencils.

5 Figure 10.16 *(right)*. Looking east at the Doon from the foreshore just north of the 5th Tee. Note the attitude of the cooling columns in the Doon stack compared with the main cliff. Also, the groundmass composition of the low stack on the left of the photo actually changes towards the left-hand side of the photograph. The group is standing on and obscuring a number of dykes.

Exercise 10: field notes exact location	General location: Date:	data

Exercise 10: field notes exact location	General location: Date:	data

Exercise 10: field notes exact location	General location: Date:	data

Part 4

Practical field skills: quick reference section

This section includes explanations and guidelines for 25 basic geological field skills. They are numbered and listed in order of their appearance during exercises 1 to 5.

Wherever you are standing in Britain your position can be described accurately to within a ten metre square by giving an 8-figure grid reference. How to do this is outlined on the side of all 1: 50 000 'Landranger' & 1: 25 000 'Outdoor Leisure' Ordnance Survey maps.

To decide where you are on the ground, either to take a grid reference of your position or find a given one requires both observation and map interpretation. Here are four basic steps:

1. Find North using the compass; in which direction does the magnetic needle point?

2. Look around for at least two obvious features of the landscape that might appear on a map, take a compass bearing to each of them and estimate their distance away.

To take a compass bearing...

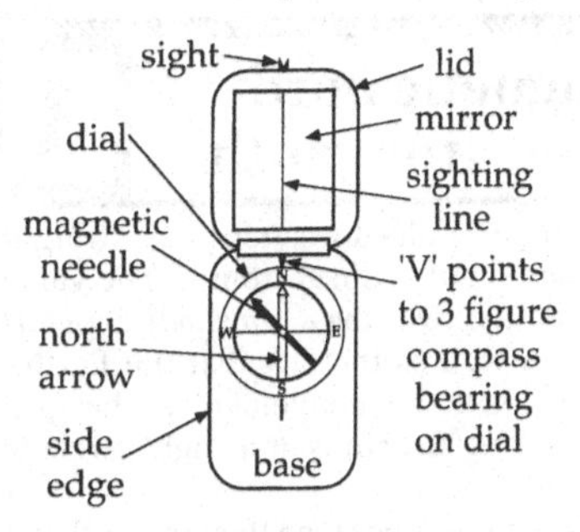

For instance, what is the compass bearing to the erupting volcano?

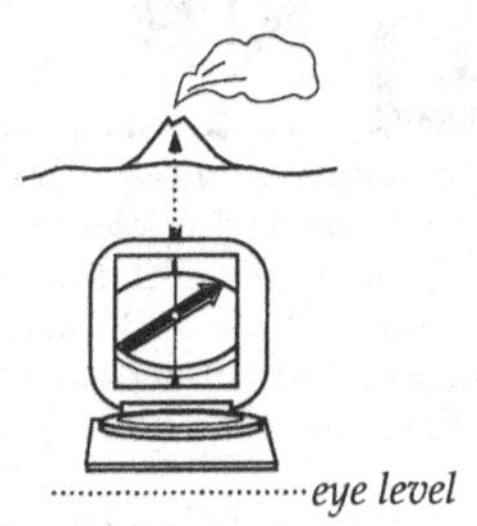

(a) Keep base of compass horizontal & bring it up to eye level
(b) Tilt lid (at ~45°) to show reflection of dial clearly in mirror
(c) Use the sighting line to point compass directly at volcano
(d) Turn dial until magnetic needle sits within north arrow in mirror
(e) Now look down on base & read off 3-figure bearing at 'V' on dial

3. Now locate one of the features on the map. Place the compass with the bearing still correctly shown on the dial onto the map and align the compass so that the **north arrow** is parallel to the grid lines that run N-S across the map. Whilst keeping these lines parallel, move the compass until one of the two side edges runs through the feature. Your position should be somewhere along this line.

4. Repeat this process for the second feature. Where the two lines intersect should be your exact position.

Always keep the magnifying lens about 3-4 cm away from your dominant eye & move the object to be studied in & out of focus with the other hand, or move your head & hand together to scan over the rock face.

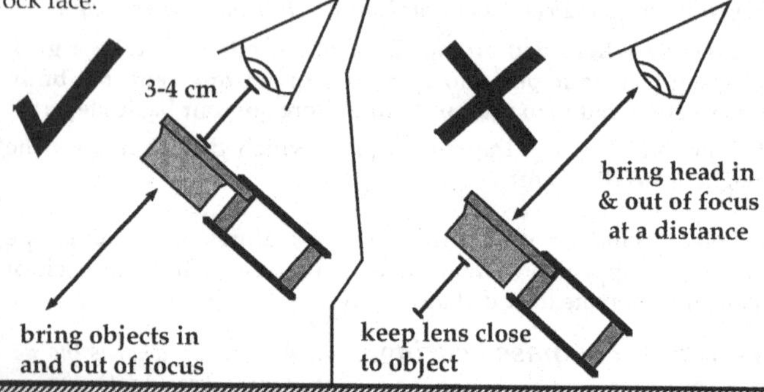

1 NB Magnetic north versus true north 1

Unfortunately, the Earth's magnetic pole does not pass straight through the north pole, or *true north*, but is offset from it. The value of this angular difference between the two (the *declination*) depends on your latitude and longitude as well as the fact that the Earth's magnetic field moves with time. Rather inconveniently, the grid north of the reference system covering Britain is at a slight angle to true north too.

It is important to bear these facts in mind because they mean that a compass and a map each point to a different north! In other words, a compass bearing from *magnetic* north to a particular landmark is going to be a few degrees different than the bearing on a map from *grid* north to the same landmark.

Luckily, the angle between the two norths on Arran is very small and it is always stated around the edge of Ordnance Survey maps wherever you are in Britain. The 1: 25 000 map of Arran quotes a value of magnetic north being 4.5° west of grid north in 1998, decreasing by 0.5° east every 4 years.

This small manual correction needs to be made to a compass bearing when it is to be used on a map to find your position, or vice versa.

The simplest way of remembering which way to make this ±4.5° correction when converting from one to the other is to use this annoying rhyme: ***Grid to Mag. add; Mag. to Grid get rid!'***

Whenever you see a new outcrop of rock, walk around it, over it, & then along it before standing back and taking stock of the general structure. Then follow the guidelines in this flow chart.

Is the rock all the same colour, texture and form, or does it look like there may be more than one lithology? → weathering will often help your eye pick out subtle lithological differences by changes in colour, texture or 'blockiness'

Are there any regular planar features running through the rock? → if there are, these may be *bedding*, *cleavage*, *jointing* (or possibly *faulting*); a closer inspection is needed to establish how many of these are present

Are there any other unusual or irregular planar features that strike you about the outcrop? → these might be due to *folding*, *faulting*, *hydrothermal veins* or *igneous intrusions*

bear all this in mind as you step up to the outcrop for a closer look

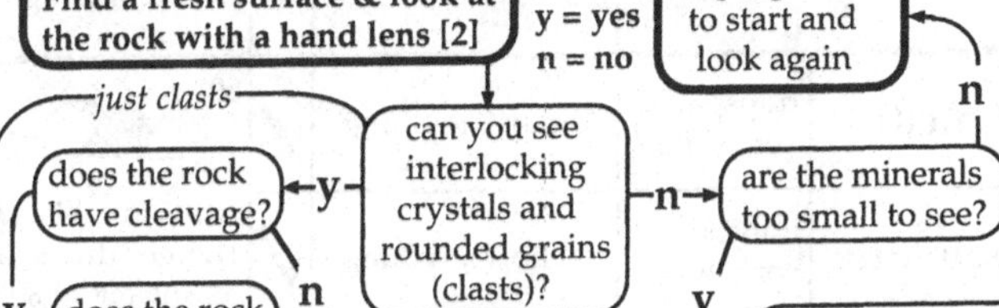

Find a fresh surface & look at the rock with a hand lens [2]
y = yes
n = no
oops, go back to start and look again
just clasts
does the rock have cleavage?
can you see interlocking crystals and rounded grains (clasts)?
are the minerals too small to see?

just crystals

does the rock have crystals in bands?

will the rock powder when scratched with metal?

does the rock fizz in 10% HCl?

it is a limestone

is it made of glass?

it is igneous

it is a fine-grained siliciclastic sediment, dolomite or gypsum

it is a clastic sediment

it is metamorphic

Now identify the main constituent minerals **[4]**, *give the rock a name* **[5]**, **[7]** *or* **[16]** *& work out how the outcrop formed*

mineral	typical colour	typical habit	scratch with metal blade?	fracture on a fresh surface
Quartz	transparent to white	prismatic crystals or granular aggregates	✗	angular & conchoidal due to lack of cleavage
Olivine	olive green	granular aggregates	✗	
Garnet	red-brown to purple	rhombododecahedron crystals	✗	
Plagioclase feldspars	white to transparent	tabular/prismatic or massive	✗	90° blocky 'steps' from 2 cleavages intersecting at 90°
Alkali feldspars	white to pink	as for plagioclase + twinning (see **[5]**)	✗	
Pyroxene	black or green-black	prismatic/massive or granular	✗	
Amphibole	black or dark green	prismatic/massive or granular	✗	56° sloping 'steps' from 2 cleavages at 56°
Calcite	transparent or white	almost anything	✓	60° splits into blocky rhombs due to 3 cleavages at 60°
Dolomite	transparent or white	almost anything	✓	
Biotite	black to dark brown	tabular prisms	✓	splits into thin flakes due to 1 very good basal cleavage
Muscovite	transparent to pale grey	platy	✓	
Chlorite & Serpentine	green to ochre	platy (serpentine has a waxy lustre)	✓	
Baryte	white to transparent	tabular in a 'cockscomb' mass *(twice as dense as the other minerals)*	✓	78° blocky fans from 2 cleavages at 78°

Select a horizontal line in chart below that describes the correct % mineral abundances; use crystal size to name rock

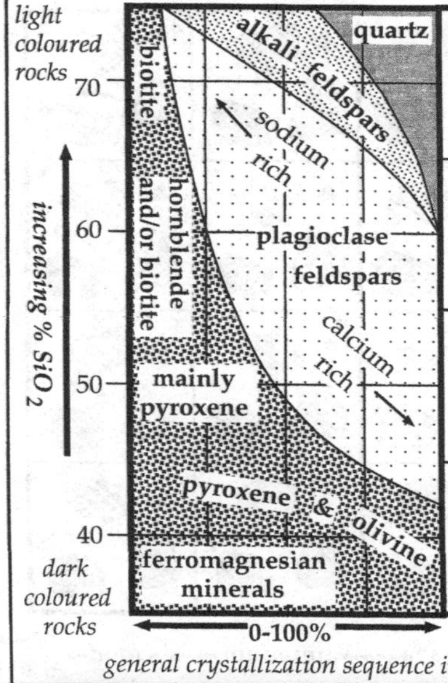

		coarse grained (≥2 mm)	*medium grained* (0.5-2 mm)	*fine grained* (≤0.5 mm)	*glass*
light coloured rocks 70	**Acid**	**granite** *(quartz conspicuous)*	**micro-granite**	**rhyolite** *(light grey or green)*	**felsite** / **pitchstone** *(black with conchoidal fracture)*
		granodiorite	**micro-granodiorite**	**rhyo-dacite**	
60	**Intermediate**	**diorite** *(hornblende conspicuous)*	**micro-diorite**	**andesite** *(paler than basalt)*	
50	**Basic**	**gabbro** *(pyroxenes conspicuous)*	**dolerite** *(black with light flecks)*	**basalt** *(black)*	
40 *dark coloured rocks*	**Ultrabasic**	**peridotite** *(olivine conspicuous)*			

increasing % SiO_2

Mineral abundance chart: quartz; alkali feldspars; sodium rich; plagioclase feldspars; calcium rich; biotite; hornblende and/or biotite; mainly pyroxene; pyroxene & olivine; ferromagnesian minerals

0-100%

general crystallization sequence in a melt:

additional rock textures

porphyritic – 'phenocrysts' (larger crystals) in a 'groundmass'

xenolithic – 'foreign' blocks of rock in igneous groundmass

'overgrowths' around 'xenocrysts' of crystals 'foreign' to the melt composition

additional crystal textures

euhedral

anhedral

zoned

simple twinning

Five principle physical properties of grains in clastic sediment need to be described: their **sorting & maturity**, **roundness**, **sphericity** & **size grade**. The figures show the differences between these five and how to describe them.

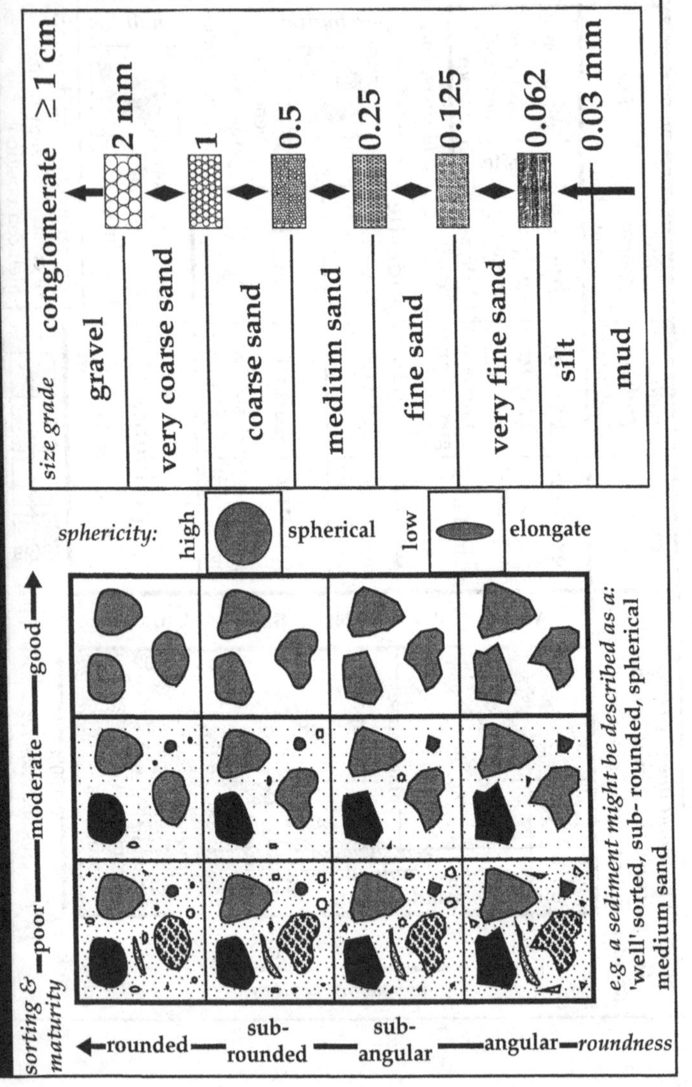

Field skill [6] demonstrates how to decribe the shape & assortment of grains in a sediment, but these physical characteristics say nothing of the mineralogical composition of a clastic sedimentary rock. The diagram below shows how to classify them.

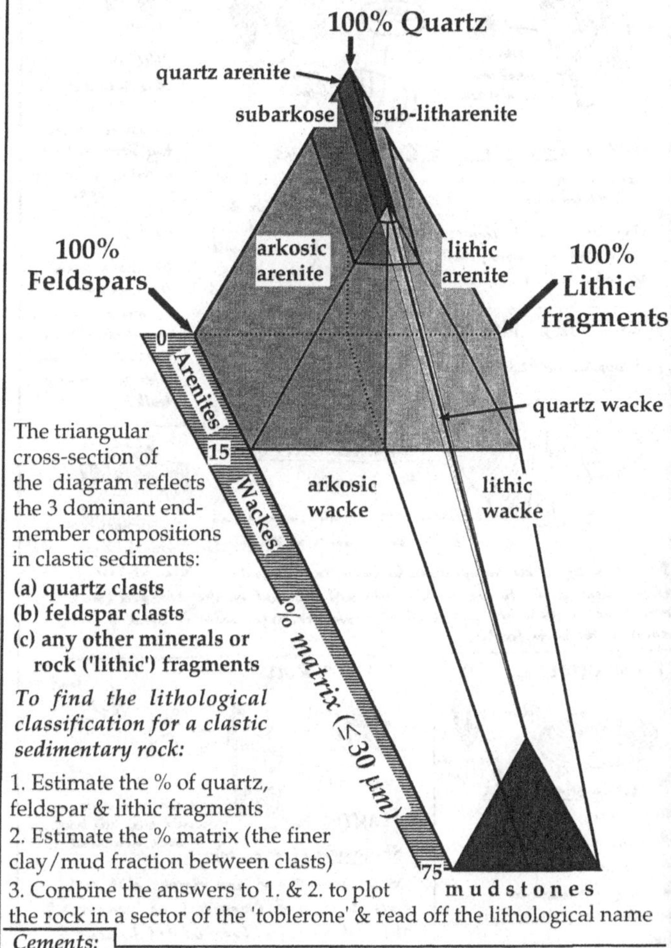

The triangular cross-section of the diagram reflects the 3 dominant end-member compositions in clastic sediments:

(a) quartz clasts
(b) feldspar clasts
(c) any other minerals or rock ('lithic') fragments

To find the lithological classification for a clastic sedimentary rock:

1. Estimate the % of quartz, feldspar & lithic fragments
2. Estimate the % matrix (the finer clay/mud fraction between clasts)
3. Combine the answers to 1. & 2. to plot the rock in a sector of the 'toblerone' & read off the lithological name

Cements:

Haematite can coat grains to make them brick red in colour.
Calcium carbonate frequently precipitates in pore spaces as cement.

OPEN MARINE (WARM SHALLOW SHELF)

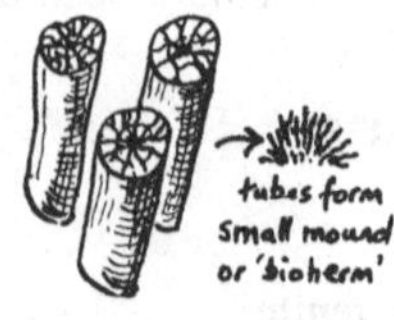

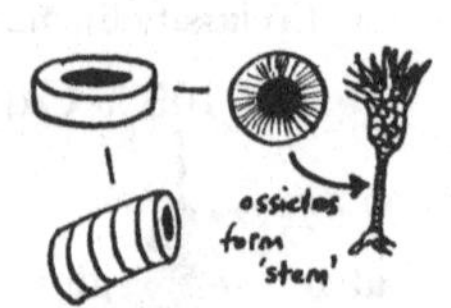

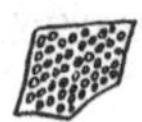

'Net' Bryozoans

Size: individual pits < 1mm.

Occurrence: these tiny organisms live in encrusting colonies on the sea-bed.

CORALS: Lithostrotion

Size: ~0·5cm in width
~4cm long

Occurrence: Corals require warm (tropical), clear, shallow seawater to thrive.

Crinoid ossicles

size: from ~3mm up to 2cm

Occurrence: These are the 'stems' of 'sea lilies' which are related to echinoids (sea urchins). They attach these 'stems' to the sea bed

Echinoid spines

size: <0·5cm

Occurrence: sea-urchins are covered in these spines which detach after the organism's death.

RESTRICTED MARINE

(LAGOONAL OR ESTUARINE)

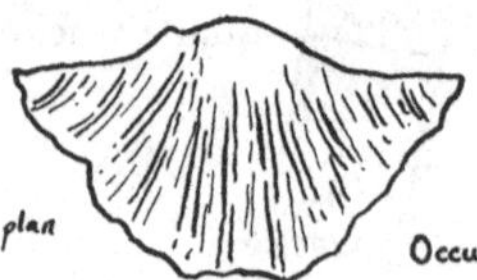

Gigantoproductus

Size: ≫5cm

Occurrence: Both these belong to the same family of Brachiopods.

Dictyoclostus

size: ~1-3cm

Both have the same adaptations to their environment; thick basal values to anchor them into soft sediment on the sea-bed (some even have spines to help with this). The smaller, upper valve is lifted up to draw water in to feed.

FRESHWATER

BIVALVES: Anthracosia

Size: 1-2cm Occurrence: A burrowing bivalve from its elongate shape. Found in only freshwater environments; rivers or lakes

TERRESTRIAL

PLANTS:

Stigmaria (tree roots)

Occurrence: These are simple 'club' roots from Lycopod trees such as 'horse-tails'. The root has often been infilled by sand & then decayed to leave only an imprint.

Sedimentary bedding can take an almost infinite variety of forms. However, a few common basic types can be recognised & these are outlined below. Often they will occur in various combinations.

Below are four basic forms that sedimentary bedding might take during periods of continual deposition, albeit at variable rates:

(i) uniform, 'massive' beds

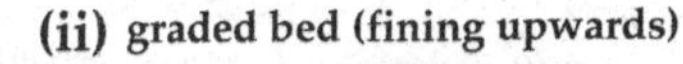

(ii) graded bed (fining upwards)

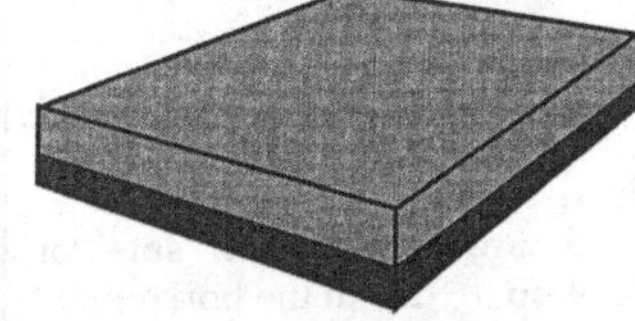

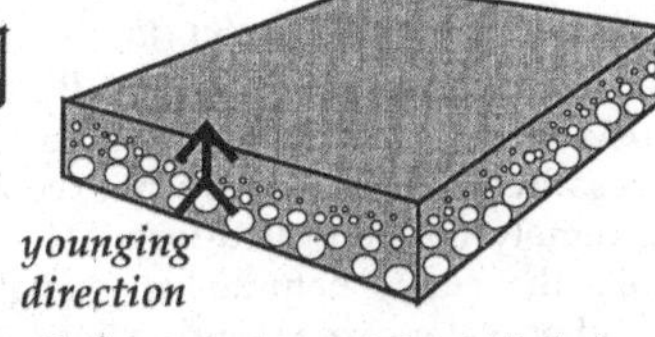

(iii) thinly laminated bed

(iv) shale (an original 'flaky' fabric to the sediment)

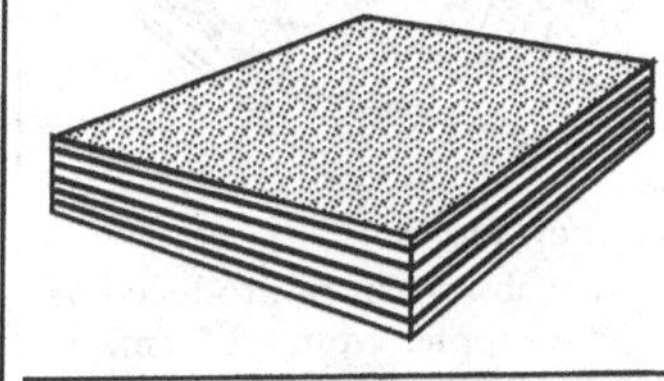

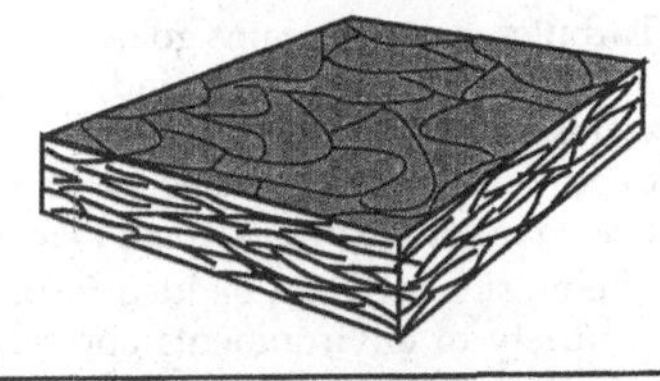

Bed-forms involving minor breaks in sedimentation (non-sequences)

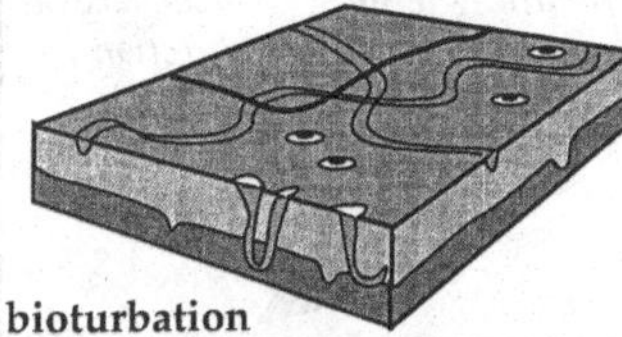

bioturbation (vertical & horizontal burrows)

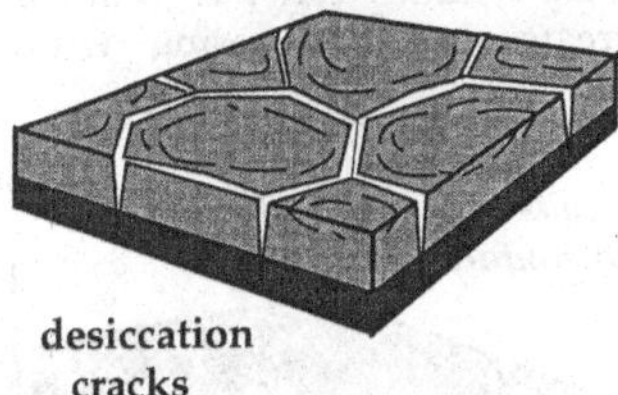

desiccation cracks

(both these suggest relatively short breaks in deposition otherwise the features would be eroded before they could be buried and preserved)

Bed-forms involving breaks in sedimentation (non-sequences or disconformities)

(broken shell material will often be swept in by water currents to fill small hollows on erosion surfaces. The time gap represented by this feature can be highly variable)

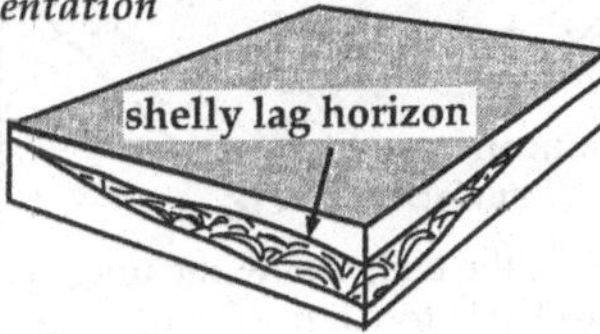

cross-bedding

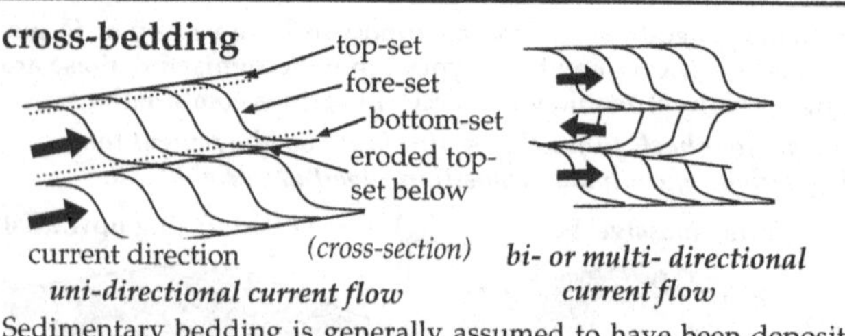

Sedimentary bedding is generally assumed to have been deposited horizontally unless there is good evidence to the contrary. **Cross-bedding** is one such exception to this rule and it is an extremely common bed-form. The beds are deposited in '**sets**', or units, that can lie naturally at an angle of up to ~**25°** to the horizontal. Wind or water carries sand grains scouring up the proximal side of a dune or ripple eroding the surface below. Turbulence as the grains go over the crest causes a loss in wind/water energy & the grains are deposited. They 'avalanche' down the distal face of the dune/ripple to make a new fore-set.

The most basic cross-bedded form shown above can be produced in a variety of environments and sizes; e.g. ripples from ~2-5 cm, or ~2-5 m dunes in rivers, strong tidal currents, or windy deserts.

Desert dunes can have curved crests due to veering wind directions.

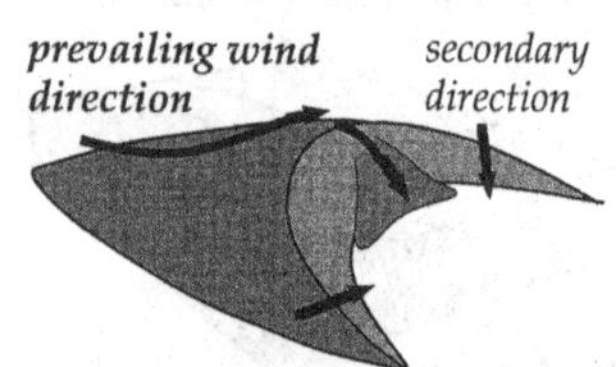

Bed-forms involving major breaks in sedimentation (unconformities)

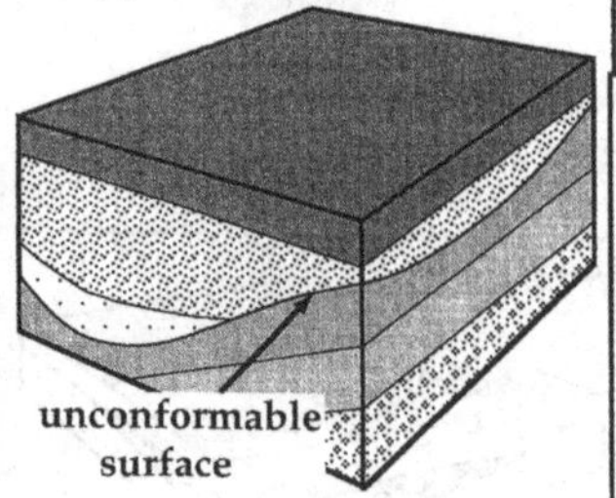

Here the beds below the unconformity have been deformed and then eroded before those on top were deposited.

Post-depositional bed-forms

Sometimes water can be held in sediment during burial. If the water is suddenly allowed to escape, this 'de-watering' can produce random, **convolute bedding**.

There are many planar surfaces or structures on them that a geologist may come across in the field: ***sedimentary bedding, metamorphic cleavage, fold limbs, dyke*** *or* ***sill margins*** *and* ***mineral veins*** *are but a few examples of them.*

In order to investigate how these structures may relate or otherwise to each other it is important to accurately describe their attitude in 3-D. This is achieved by measuring two different 2-D angles, and then combining them together.

The first is the angle between the surface (or structure) and horizontal. It is termed the **angle of dip** and is measured using an instrument called a **clinometer.**

The second is the angle between **north** and *either* the direction in which the surface is dipping (the **dip direction**), *or* at right angles to this dip direction (the **strike**). This angle is a simple three figure bearing and so is measured using an ordinary compass (see **[1]**).

Using a clinometer

Combined compass-clinometers are available & a serious student of geology should purchase one. However, they are relatively expensive (about £50) and an effective, though obviously less accurate clinometer can be made for about £1. This must then be used in conjunction with an ordinary compass (which can be ≤ £10).

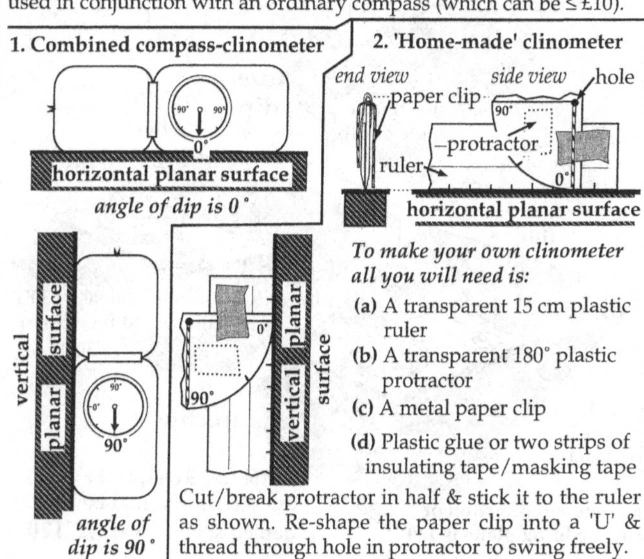

To make your own clinometer all you will need is:

(a) A transparent 15 cm plastic ruler

(b) A transparent 180° plastic protractor

(c) A metal paper clip

(d) Plastic glue or two strips of insulating tape/masking tape

Cut/break protractor in half & stick it to the ruler as shown. Re-shape the paper clip into a 'U' & thread through hole in protractor to swing freely.

To measure the dip & dip direction of a planar surface:

(i) Find the maximum angle of dip of the surface (α). Drop water onto the surface or spit on it; both will trickle down the steepest slope and leave a line along it.

(ii) Place the edge of the clinometer on this line and hence measure the angle of dip of this surface from the horizontal.

(iii) Place the rear edge of the compass flat against the planar surface and keep it there whilst twisting the compass around until it is held level & horizontal. (A spirit level stuck to the compass base-plate is used by many geologists for this purpose. Alternatively, tap the dial with a finger as you move the compass; if the **magnetic needle** wobbles freely then the compass is approximately horizontal).
Take the bearing of the direction in which the compass is pointing (see [1]); this is the **dip direction.**

(iv) Alternatively, the **strike** (at 90° to dip direction) can be measured.

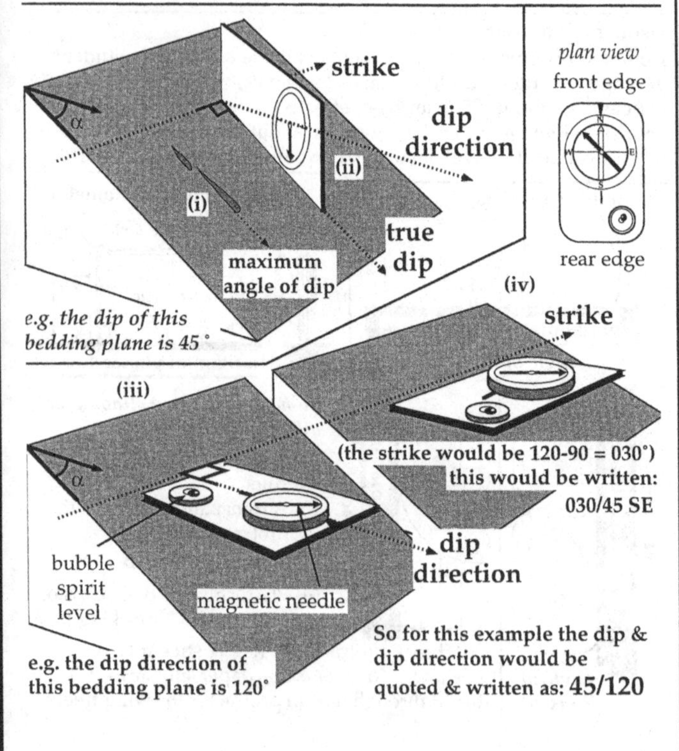

When collecting dip and dip direction measurements from planar structures such as bedding it is often useful to plot the data graphically to see if there is any preferred orientation to them. Commonly, this is done in one of two ways; either by using a **rose diagram**, or a **stereonet** (for another use of a stereonet see **[19]**).

The **rose diagram** is the simpler of the two methods and is essentially a cumulative frequency histogram curled round into a circle to represent the degrees of a compass. The circular **x-axis** is divided from 0° to 360° and it is on this that the dip direction is plotted. The radiating **y-axis** is the cumulative number of times each dip direction is recorded; here on a scale of 0 to 9.

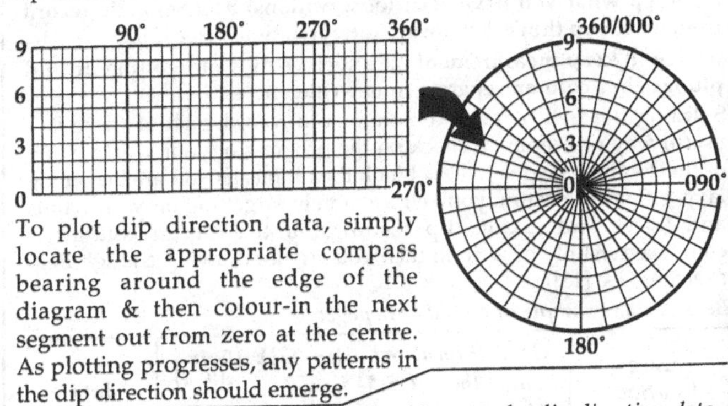

To plot dip direction data, simply locate the appropriate compass bearing around the edge of the diagram & then colour-in the next segment out from zero at the centre. As plotting progresses, any patterns in the dip direction should emerge.

360/000°
9
6
3
0
'A'
'B'
270°
090°
180°

e.g. In this example, dip direction data were collected from the bedding planes of two sedimentary units; 'A' & 'B'. Unit 'A' underlies 'B'.

The data have been plotted on the rose diagram and shaded black. The measurements clearly show an angular difference in dip directions between the two sedimentary units. This can be interpreted as an angular unconformity separating the two.

Rose diagrams do not have to be used just for plotting dip direction data, but can be applied to any geological problem where compass bearing measurements are taken. For instance, the preferred orientation of currents as shown by ripples, sand dunes or even fossils.

*The time spent making field observations, interpretations & collecting data is wasted unless you record them either for later use by yourself or for others to follow the geological descriptions that you make (see **Exercise 3**).*

It is always remarkable how quickly you can forget the finer details of outcrops, or worse still, get observations from different outcrops jumbled up in your mind. So you **must record what you see when you see it**.

There are no hard and fast rules over how to write field notes, but here are some guidelines:

(a) Always record your exact location (with grid references) before writing anything else (see [1]).

(b) Keep what you have seen (observations) as a separate record from what you think they mean (interpretations) (see [3]).

(c) Try to keep measurement data such as dips of bedding or fold plunges in a column separate from written notes.

Splitting the field notes into columns to record each of these is a useful way of keeping a check on the progress of your work. If any of the columns begins to look blank then it means you are becoming lazy, either in locating yourself accurately or getting on your hands and knees and taking dip readings. If the central column is composed mainly of writing then you are not drawing enough field sketches (see [13]).

Below is an example of a field note page:

exact location	*General location:* Glen Tilt (Blair Atholl) *Date:* Thurs 2-5-96 a.m. *Sheet No:* 1	*data*
	Left car at Gilbert's Bridge & walked to Clachghlas (1hr 15 mins).	
Grid 919 729 ~100m upstream from wooden bridge at Clachghlas - outcrops along waters edge on river bank	- All along this stretch of the Tilt the NW bank exposes two lithologies: ⓐ Red alkali feldspar ~50%, ~2-4mm White (?Plag or alkali) feldspar ~30%, ~2mm grey quartz ~20%, ~2-4mm = Red Granite	Granite-schist contacts have variable dip as granite 'veins' the schist
	ⓑ Quartz grains & crystals ≥50%, of variable size Muscovite ~30%, >2mm Other minerals (? feldspars, clays) ≤20% as matrix. A pervasive planar fabric runs through this lithology discordant with lithological variations (a metamorphic cleavage) = Quartz-mica schist	Regional dip of Schists: 80/196 74/192 86/188

Sketching is an indispensable field skill. It condenses and summarises the important geological features of an outcrop and can save much time in trying to describe it in words. Sketches need not try and rival the French 'Impressionist' painters, but instead the emphasis should be in getting as much geological information as possible onto an accurate, but simple, outline.

Any sketch needs to be accompanied by 3 pieces of information:

(a) The exact location (a grid reference + relation to a nearby obvious landmark)
(b) The direction in which you are looking to draw the sketch (a 3-figure compass bearing)
(c) Time of day & approximate state of the tide (if coastal) or vegetation (in summer many outcrops can be covered).

In a similar way to field notes, there are no hard and fast rules on how to draw a field sketch. However, below are some guidelines for producing an accurate sketch containing the maximum information.

1. Imagine a frame around the area to be sketched & draw this on the page: the size of this frame will depend on what features need to be in the sketch & what can be left out.

Then construct a feint grid within the frame on the page.

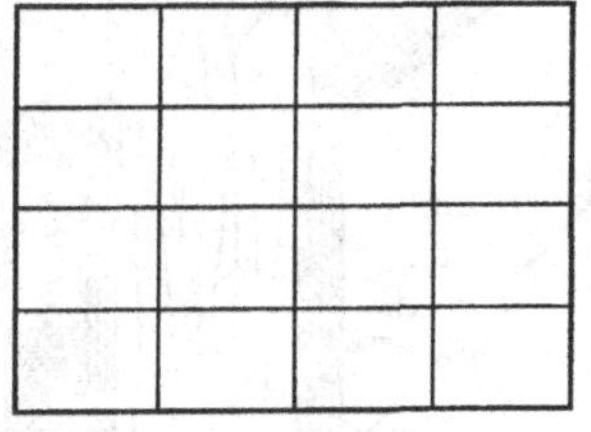

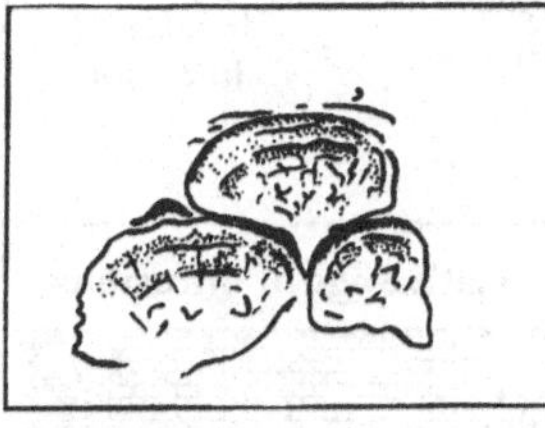

2. Now sketch the outline of the outcrop & the important geological features within it. Use the grid to make sure you draw everything in proportion, at its correct size & at the right attitude to the rest of the outcrop.

3. Estimate a scale for the sketch & draw a scale bar underneath it. If sketching at a distance use people, trees or anything that might give an idea of the size of outcrop & draw them in.

Finish by highlighting & labelling the geology

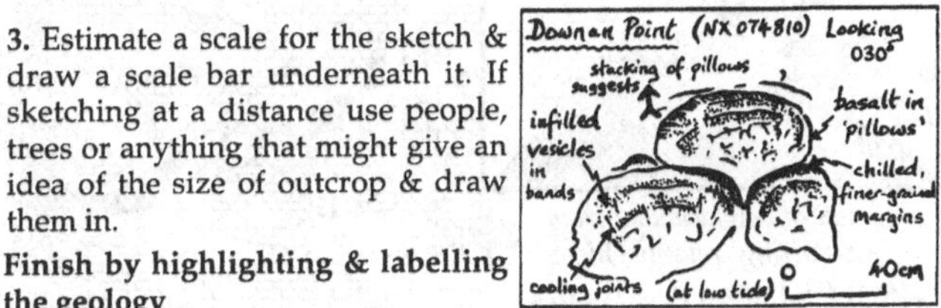

(i) Emplacement structures: during intrusion of acidic melts where the viscosity is high, shear planes can be set up within the flowing liquid which can give rise to **flow-banding**, or even **flow-folding**.
(ii) Cooling structures: polygonal/**columnar** cooling **joints** can be formed as the melt cools, crystallises & 'cracks' as it contracts. These joints always form **perpendicular** to the cooling surface.
(iii) Intrusion orientation: this should always be measured as a **trend** in dykes (a compass bearing) & as the **dip & dip direction** of the edges (**contacts**) of the intrusion in both dykes & sills.
(iv) Relative age of intrusions: found by determining which younger intrusions have off-set or '**cross-cut**' older ones during emplacement.

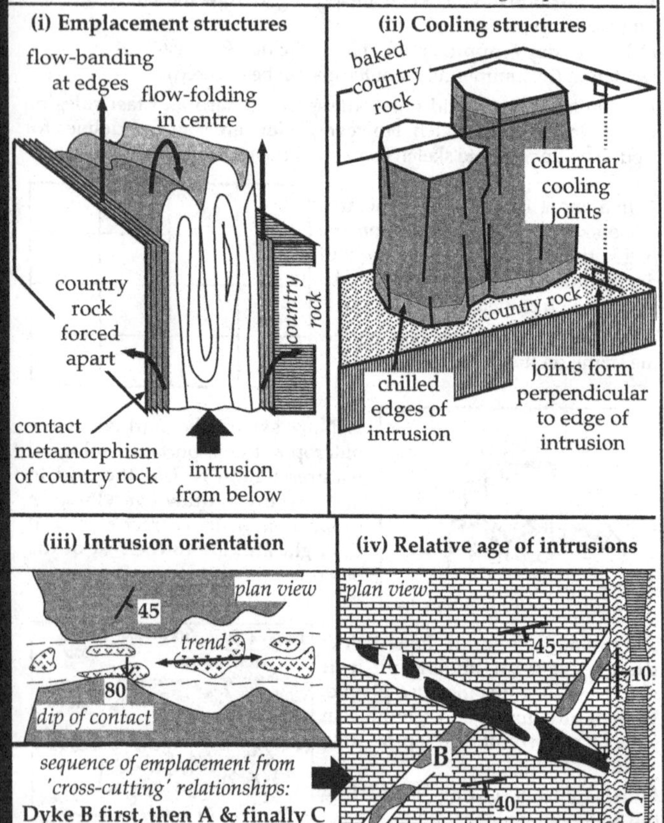

Cleavage

Metamorphic cleavage is a planar fabric that develops in rocks due to various types of deformation, often cross-cutting any sedimentary bedding. The five most common forms are outlined below.

1. Slaty cleavage

This is a penetrative, repeating cleavage that develops mainly in fine-grained, incompetant units (see [16]).

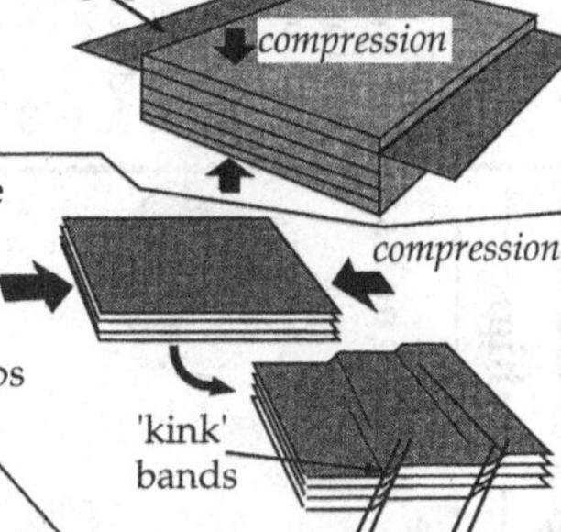

2. Crenulation cleavage

This is a repetitive micro-folding developed in an already existing planar foliation such as **slaty cleavage**. The mechanism is perhaps best imagined as corrugating iron.

3. Pressure-solution cleavage

heat & compression

This is a lithological banding that develops in a moderately high temperature & pressure metamorphism where it leads to actual mineral segregation and migration within the rock.

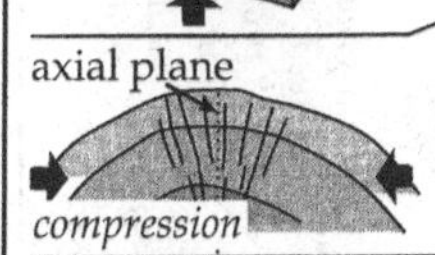

4. Axial-planar cleavage

Compression and fracturing along the axial plane of a fold can lead to a localised cleavage developing parallel to the axial plane in the hinge zone (see [17]).

5. Fracture cleavage

This is a localised deformation feature usually associated with faults & is a series of closely spaced, small-scale fracture planes.

Jointing

This is another secondary type of planar fabric that rocks can develop. It differs from cleavage in that it is not caused by metamorphism as such, but by brittle fracture of the rock, commonly through expansion.

Dilation (extension) joints: These are extremely common and can often be the most obvious planar fabric in outcrops, forming joint 'sets' at about 90° to each other. They can form due to expansion from unloading as sediment is eroded away above the strata, or from folding competent rocks.

Shear joints: These can be produced if the rocks undergo strain & there is limited movement along the the opening joints (see [17])

Cooling joints: See [14]

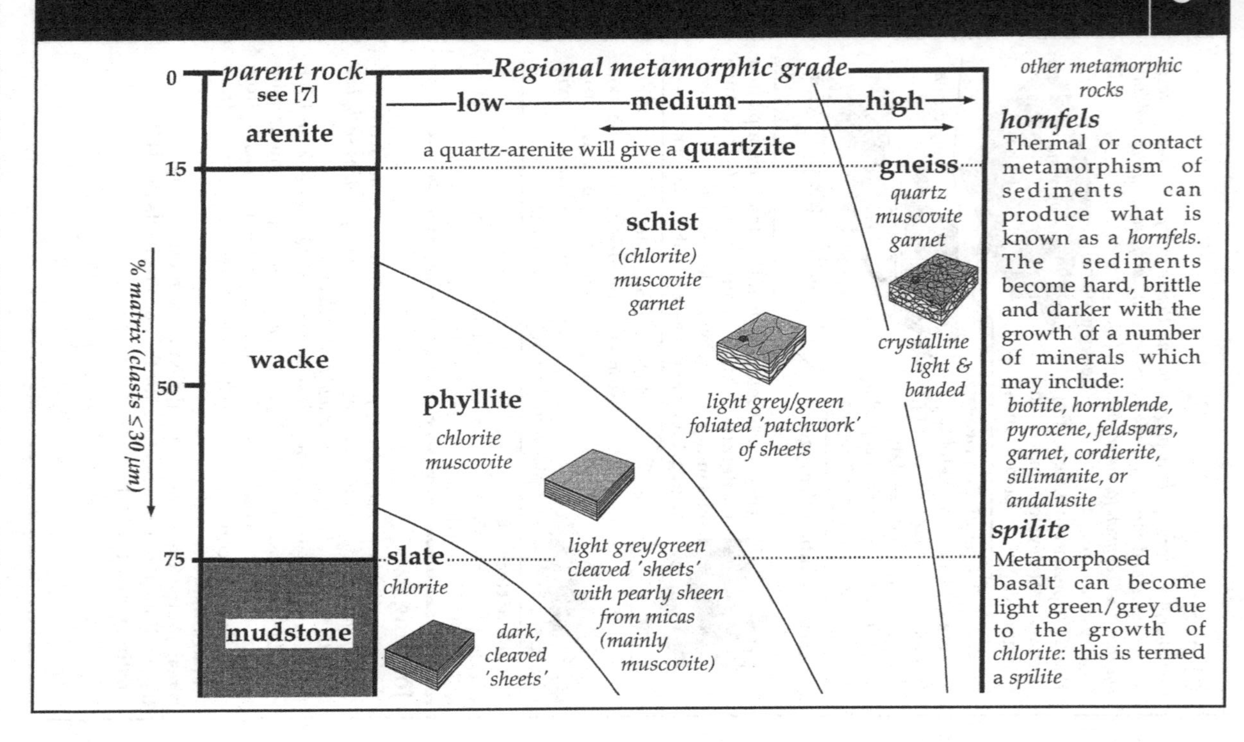

parent rock
see [7]
Regional metamorphic grade
low
medium
high
other metamorphic rocks
0
15
50
75
% matrix (clasts ≤ 30 µm)
arenite
wacke
mudstone
a quartz-arenite will give a quartzite
gneiss
quartz
muscovite
garnet
crystalline
light &
banded
schist
(chlorite)
muscovite
garnet
light grey/green
foliated 'patchwork'
of sheets
phyllite
chlorite
muscovite
light grey/green
cleaved 'sheets'
with pearly sheen
from micas
(mainly
muscovite)
slate
chlorite
dark,
cleaved
'sheets'
hornfels
Thermal or contact metamorphism of sediments can produce what is known as a hornfels. The sediments become hard, brittle and darker with the growth of a number of minerals which may include:
biotite, hornblende, pyroxene, feldspars, garnet, cordierite, sillimanite, or andalusite
spilite
Metamorphosed basalt can become light green/grey due to the growth of chlorite: this is termed a spilite

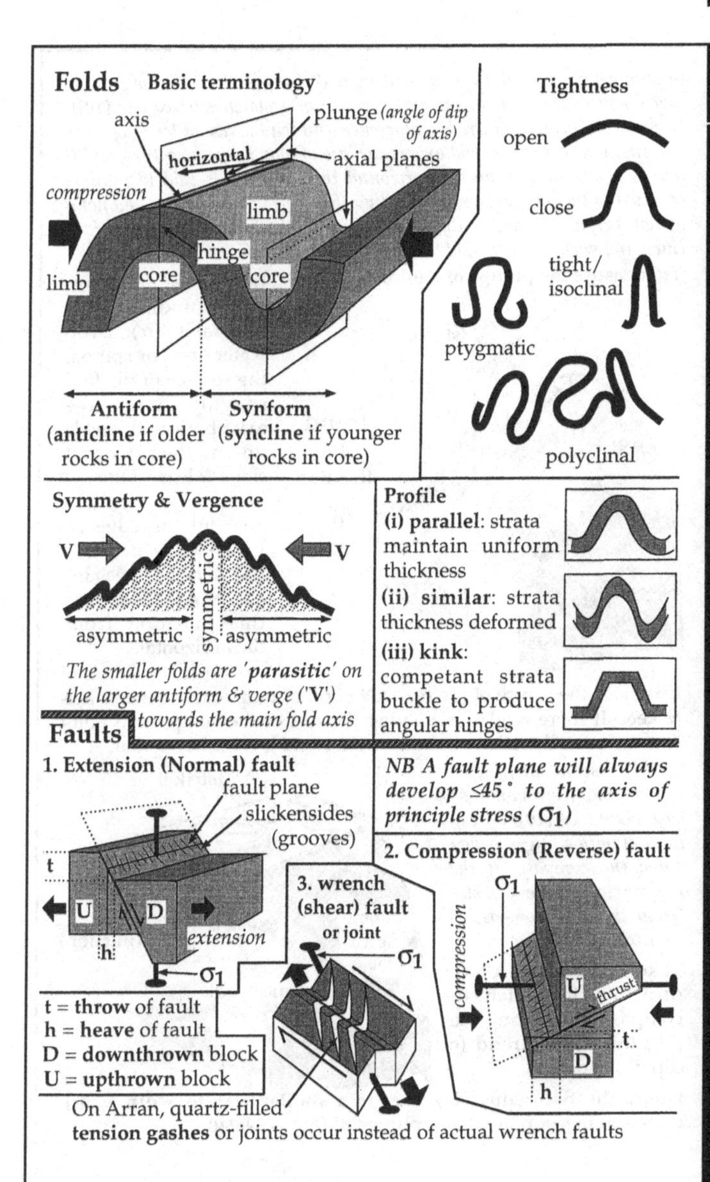
Folds Basic terminology
axis
plunge (angle of dip of axis)
horizontal
axial planes
compression
limb
hinge
limb
core
core
Antiform
(anticline if older rocks in core)
Synform
(syncline if younger rocks in core)
Tightness
open
close
tight/ isoclinal
ptygmatic
polyclinal
Symmetry & Vergence
V
V
symmetric
asymmetric
asymmetric
The smaller folds are 'parasitic' on the larger antiform & verge ('V') towards the main fold axis
Profile
(i) parallel: strata maintain uniform thickness
(ii) similar: strata thickness deformed
(iii) kink: competant strata buckle to produce angular hinges
Faults
1. Extension (Normal) fault
fault plane
slickensides (grooves)
t
U
D
extension
h
σ1
NB A fault plane will always develop ≤45° to the axis of principle stress (σ1)
2. Compression (Reverse) fault
σ1
compression
U
thrust
t
D
h
3. wrench (shear) fault or joint
σ1
t = throw of fault
h = heave of fault
D = downthrown block
U = upthrown block
On Arran, quartz-filled tension gashes or joints occur instead of actual wrench faults

In practice, the procedure for measuring fold plunges is exactly the same as that for measuring the dip & dip direction of a planar surface (see **[10]***). However, in principle there is a difference between the two in terms of what is actually measured. A fold axis is the line of maximum curvature, and if this is at an angle from the horizontal then the fold is said to* ***plunge****. Measuring this plunge means recording the angle of dip of a line and not a planar surface; consequently, a slightly different terminology is used. A* ***lineation*** *such as this is said to have a* ***trend*** *rather than a* ***dip direction****.*

To measure the plunge of a fold:

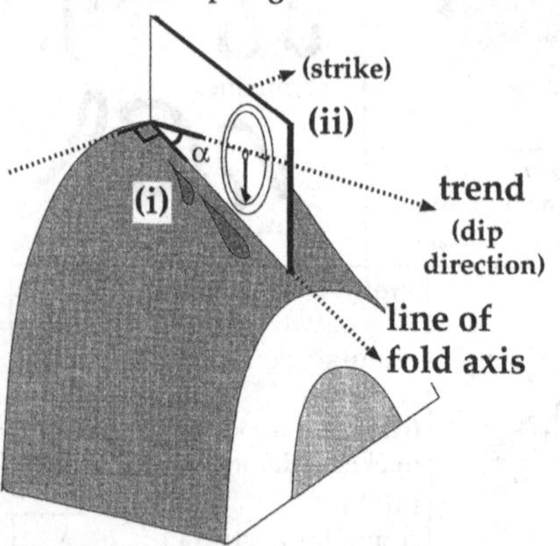

(i) Find the maximum angle of dip of the fold axis (α). Drop water onto, or spit on, the surface of the fold roughly down the axis; both will trickle down the steepest slope & leave a line.

(ii) Put the edge of the clinometer on this line & hence measure the angle of dip of this lineation from the horizontal.

(iii) Place the centre of the rear edge of the compass on the fold axis & keep it there whilst orientating the rest of the compass to point exactly in the direction of plunge whilst held level & horizontal.

(Either use a spirit level stuck to the compass base-plate (see **[10]***) or tap the dial with a finger as you move the compass; if the* ***magnetic needle*** *wobbles freely then the compass is approximately horizontal).*

Take the bearing of the direction in which the compass is pointing (see [1]); this is the **trend** (or dip direction).

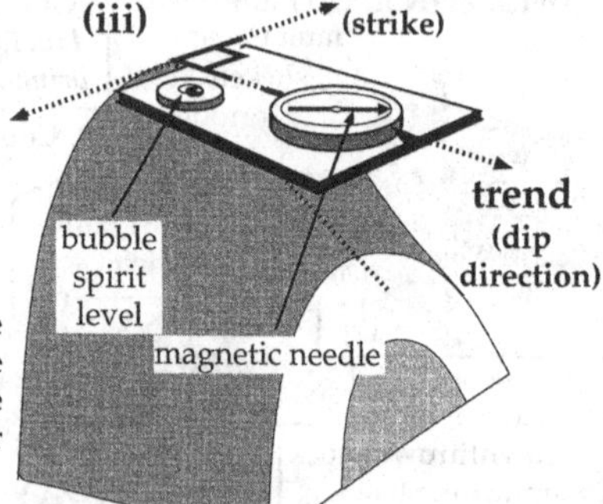

Quote the two values together in a similar way to a dip & dip direction measurement; e.g. plunge of fold is 34/240°

Initially a polar plot looks very similar to a rose diagram (see **[11]**). *However, although the lines describe a circular piece of graph paper in 2-D, they actually represent looking down into the bottom half of a sphere in 3-D.*

The usefulness of this 3-D graph, or stereonet, in geology is the ability to plot the attitude of any three-dimensional surface or line in space relative to other ones and potentially find and quantify the relationship of one to the other.

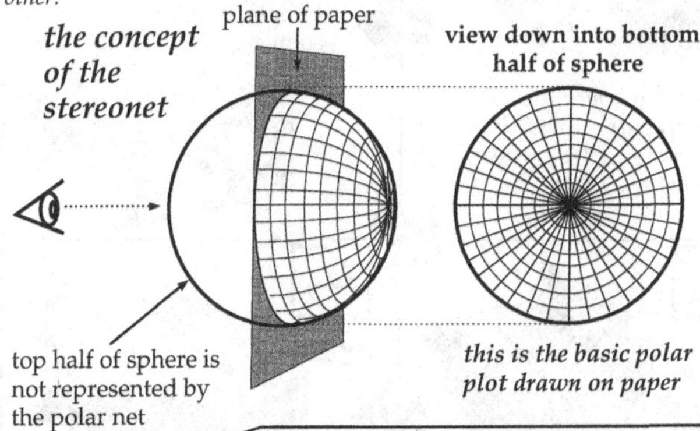

For example, where the plunge of a fold axis is measured, it will be represented on the polar plot by a single point where the line along the axis cuts the bottom of the sphere. The angles between folds with different plunges can be found from the stereonet and hence the number of deformation phases and their compression directions.

plotting the plunge of a fold axis

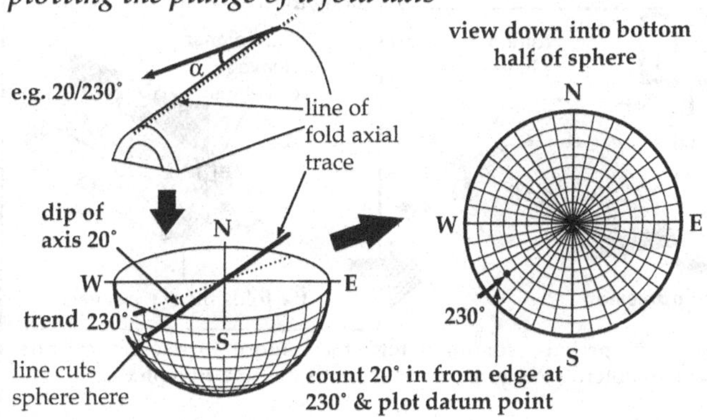

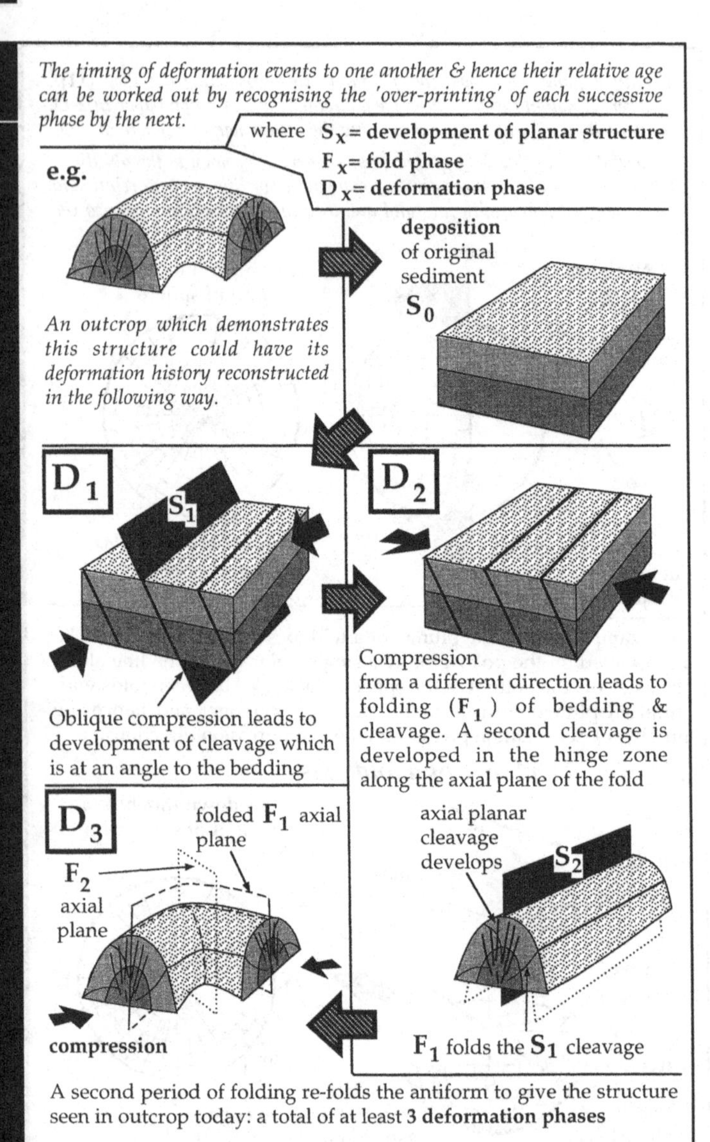

The timing of deformation events to one another & hence their relative age can be worked out by recognising the 'over-printing' of each successive phase by the next.
where S_x= development of planar structure
F_x= fold phase
D_x= deformation phase
e.g.
An outcrop which demonstrates this structure could have its deformation history reconstructed in the following way.
deposition of original sediment
S_0
D_1
S_1
Oblique compression leads to development of cleavage which is at an angle to the bedding
D_2
Compression from a different direction leads to folding (F_1) of bedding & cleavage. A second cleavage is developed in the hinge zone along the axial plane of the fold
axial planar cleavage develops
S_2
F_1 folds the S_1 cleavage
D_3
folded F_1 axial plane
F_2 axial plane
compression

A second period of folding re-folds the antiform to give the structure seen in outcrop today: a total of at least **3 deformation phases**

Basic mapping procedure

Before mapping begins

• Reconnoitre the map area first, making a note of localities that either look complicated or have no exposure in them. This can save wasting time and making mistakes later when mapping is underway.

• Attempt to establish the stratigraphy of sediments (and igneous rocks if possible) in the mapping area and draw the stratigraphic column on the side of the field map for reference.

During field mapping

• Map on the outcrops in pencil (see below), accompanied by notes, sketches and dip/dip direction measurements.

• Measure the dip/dip direction of geological boundaries whenever found & mark these on the map immediately (see [22]).

• Colour outcrops lightly in pencil to help high-light areas of the same lithology.

• Draw on the traces of fold axes & faults when found (see [23]).

• Try drawing a quick cartoon cross-section through the area (see [24]); does it make geological sense, or do some outcrops need to be revisited?

After returning from the field

• Carefully go over all the pencil lines in ink to make them permanent.

• Extrapolate all geological boundary lines, fold axes and faults.

• Lightly shade-in the areas corresponding to each lithology.

• Construct a scale cross-section to demonstrate the structure of the area (see [24]).

Drawing outcrops on a base-map

At each outcrop always locate yourself on the map [1] before continuing with observations and making notes (see [3] & [12]). Once you are satisfied with your work at an outcrop, the next step is to mark it on the map. There are several ways in which this can be done and they depend on how much space there is on the map, as well as the type of mapping that you are doing.

Draw an outline of the outcrop as accurately as you can. Think about the scale of the map; your outcrop may only be the size of a pencil point, or it could be an entire crag already shown on the map.

If your outcrop is small, it might be best to represent the outcrop by a dot or small circle. It is then given a code number or letter which can correspond to a description of it in your notes.

e.g. **. B5** **∘A1**

C3 (try not to use arrows as these might be mistaken for structural symbols)

In most cases the outcrop wil be large enough for you to draw around. A code number or letter can also be used here to refer to a more detailed description in the notes if necessary.

If the outcrop coincides with a symbol already marked on the map, such as a crag or flat-rock, then you must decide how much of the symbol is actually exposed as the outcrop and draw on its outline.

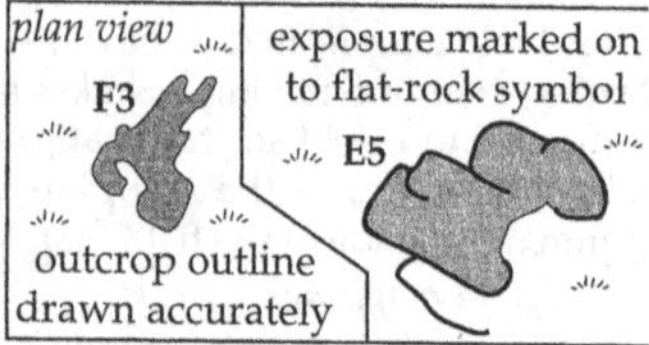

If you have soft coloured pencils then decide on a colour for the lithology of the outcrop & lightly shade it in. Remember also to put a dot of the colour next to the appropriate lithology on the stratigraphic column. Using colour on outcrops is an excellent way to see at a glance how the mapping is progressing, but you must be careful to shade gently; do not engrave the map as we all make mistakes & coloured pencil can be difficult to remove with a rubber.

Drawing structural symbols on a base-map

Once an outcrop is marked on the map it is important to accompany it with the data collected there, such as the dip of strata, geological contacts or intrusions. There are a variety of symbols that can be used on a map to represent these structures. These are summarised on the next page.

The most common symbol used is the 'bar & tick' for sedimentary bedding. This carries 4 important pieces of information.

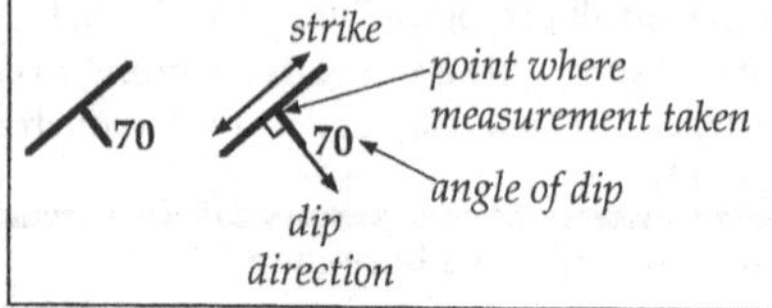

To plot a dip & strike reading accurately on a map *(see next page)*

1. Ignore the magnetic arrow on the compass and turn the dial so that the 'V' points to the measured dip direction.

2. Line-up the black lines on the base of the compass parallel with the N-S grid lines on the map. Whilst keeping these lines parallel, move the compass so that you can draw a small tick along the edge of it onto the map at the outcrop.

3. Repeat step **1.**, but this time turn the dial an extra 90° on from the dip direction. Repeat step **2.**, but draw this second line longer than the first one, making sure that the two intersect at the point of measurement and that the shorter 'tick' is the correct side of the longer strike line.

4. Finally, write the value for the angle of dip next to the 'tick'.

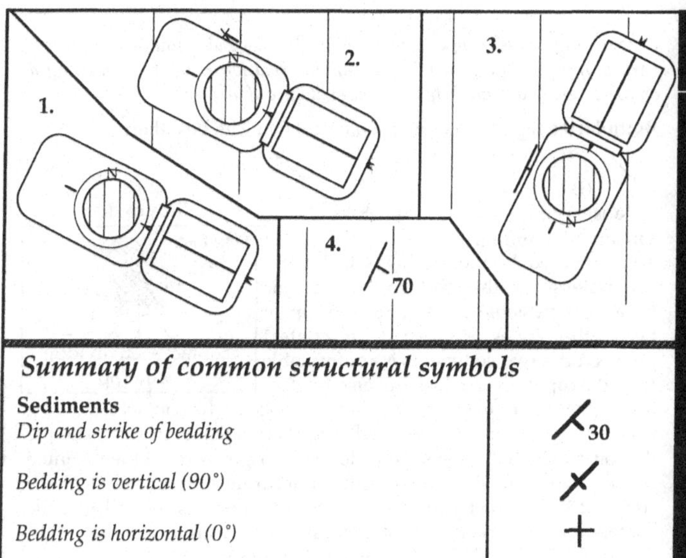

Summary of common structural symbols

Sediments *Dip and strike of bedding*	30
Bedding is vertical (90°)	
Bedding is horizontal (0°)	+
Dip and strike of bedding, but you have evidence that it is overturned	30
The beds are dipping, but you have no idea which way up they are	30
The 'younging direction'; you cannot find any bedding planes to measure but ***'way-up' structures*** *show in which direction the younger beds are. The arrow points towards the younger sediments.*	
Metasediments	
Dip and strike of metamorphic cleavage	30
Cleavage is vertical (90°)	
Igneous	
Dip and dip direction of the contact between an igneous body and the surrounding country rock, or other body	30

Recognising, measuring and marking on lithological contacts is a critical part of any geological mapping as this information is used to construct a cross-section and hence help solve the structure of an area.

Boundaries represented on a geological map can be either:

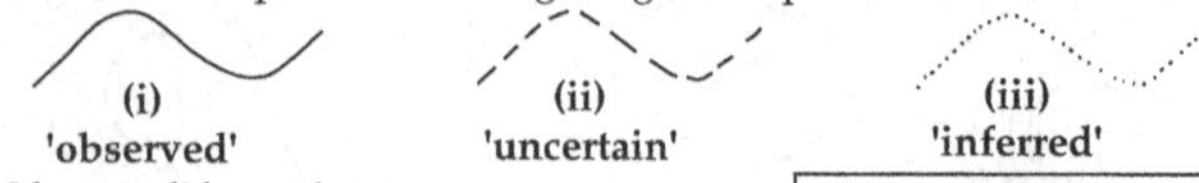

'Observed' boundary

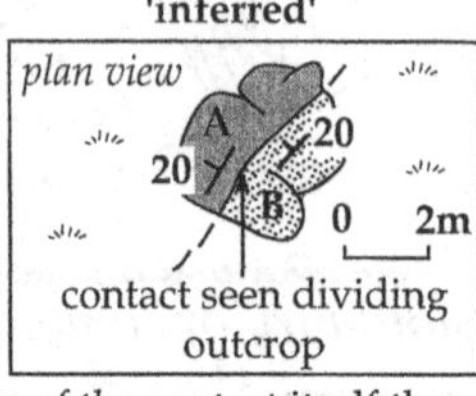

The 'observed' boundary should be hunted for tirelessly in the field. In the example here, it is possible to measure the dip of both lithology **A & B**, so if the contact between the two sediments is conformable then the dip of the contact will also be 20°. If it is possible to measure the planar surface of the contact itself then it should be done. An 'observed' boundary such as in this example demonstrates that if the sediments are the right way up then **A** must be younger than **B**. Armed with an accurate measurement of the strike of the contact you can try to follow it across the hillside. the contact is also marked easily & accurately onto a cross-section.

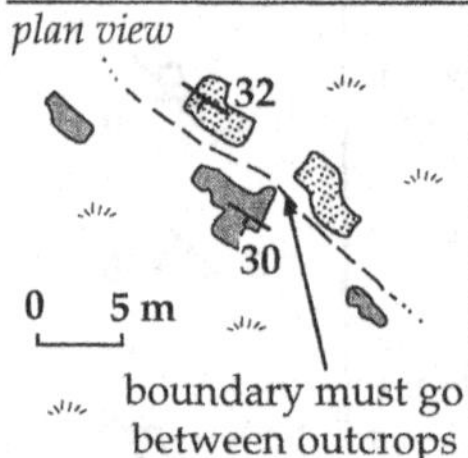

'Uncertain' boundary

An 'uncertain' boundary means that you are unable to actually place a finger on it, but using changes in lithology of outcrops & possibly a sudden break in slope or vegetation you can pin it down to within a few metres. Again, if there are plenty of outcrops around you may be able to estimate the dip & dip direction of the contact with confidence for a cross-section.

'Inferred' boundary

The 'inferred' boundary is where there is just too little exposure to be able to follow the contact with any certainty. You still need to extrapolate the course of a contact between outcrops, but you should be prepared to change this perhaps after further mapping.

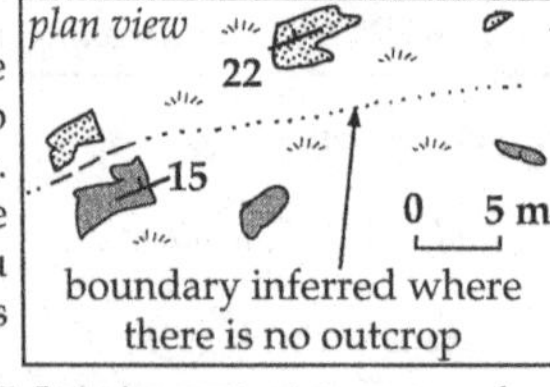

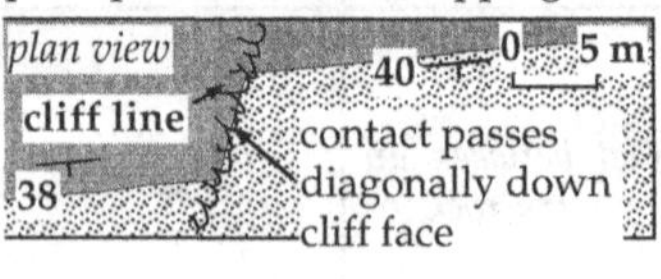

NB It is important to remember when drawing any boundary onto a map that you must take account of the topography over which the boundaries pass.

Faults

To mark the course of a fault onto a geological map, use the same basic method and symbol as for lithological boundaries (see [22]), but either use a different colour or make the line thicker.

Where the fault is observed use a solid line & put a small 'tick' on the downthrow side if you can work it out (see [17]). Also, measure the dip of the fault plane if you can. Then use a dashed line for an 'uncertain' fault & a dotted line for an 'inferred' course of a fault.

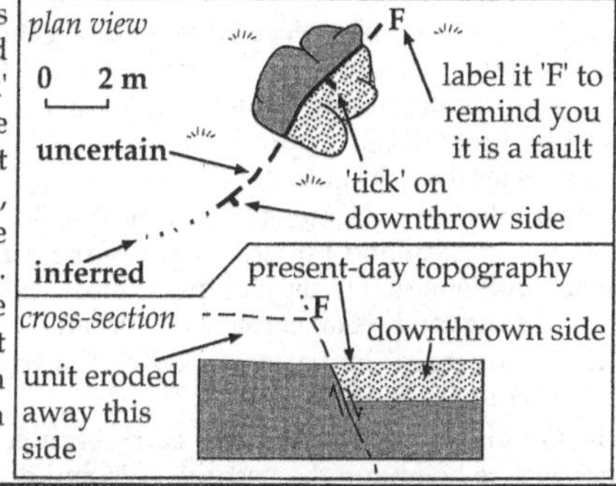

Folds

Folds may be observed, uncertain or inferred too. On a geological map they are shown by drawing on the trace of the fold axis. A small-scale fold is easy to spot in the field, but in many cases folding is on a larger-than-outcrop scale and the only way you will spot one is by a changing dip & dip direction of the strata. This is just one of the many reasons why it is important to mark dip readings onto the map whilst you are still in the field.

The line marking an axial trace on a map is shown with enclosed 'V' symbols drawn onto it to indicate the dip of limbs, so differentiating between anticlinal and synclinal folds (see [17]).

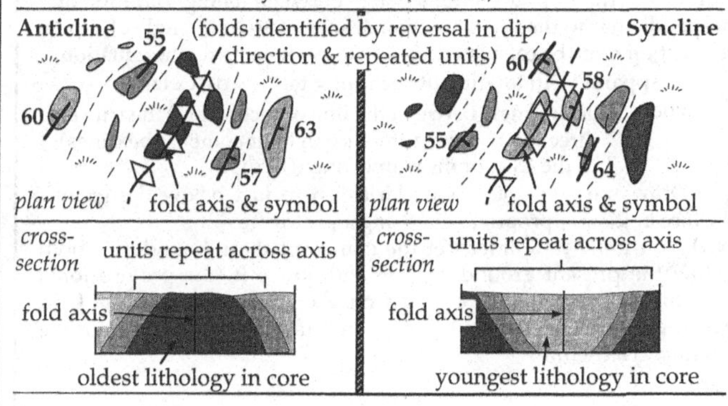

(i) Decide on the line of section across the map. It should traverse the most geologically 'interesting' area. Draw the line on and put a cross at each end of it.
(ii) On a piece of graph paper draw a horizontal line the same length as your line of section.
(iii) Place the edge of a piece of paper along the section line, mark on the end points & put 'ticks' along its edge every time it crosses a topographic contour line. Label these with the correct height value.
(iv) Place the 'ticked' paper along the baseline on the graph paper and transfer the ticks onto it.
(v) You must now decide on the vertical scale. A true section should have the vertical and horizontal scales the same. If this is likely to cause problems (such as making it cluttered or unclear) then the vertical scale can be exaggerated, but try and keep this to a minimum.
(vi) Put a vertical line at each end of the section and mark on the vertical contour intervals.
(vii) Go along the baseline and carry each contour 'tick' up to the appropriate height on the vertical scale and draw a small cross. Join the crosses to give a profile of the topography.
(viii) Repeat step **(iii)** but this time mark on the geological boundaries, faults and dykes on the paper instead of topographic contours and transfer them to the section following step **(iv)**.
(ix) Again carry the position of the boundaries up from the line and make a small 'tick' at the appropriate position on the ground surface.
(x) Now which way you draw the boundaries above and below ground depends on the angle and direction of dip that you should have measured whilst in the field.
If the **dip direction** of the rocks is **parallel** to the line of section then the angle of dip is **true** and can be marked onto the section directly. But often the dip directions vary between being parallel and perpendicular to the line of section & so an **apparent** angle of dip on the section must be calculated. This can be done using the equation;

apparent dip (on line of section) $= \tan^{-1} (\cos \mathbf{A} \times \tan \mathbf{T})$

where **A** is the angle between the line of section and the true dip direction (i.e. the difference in their compass bearings),
T is the true, or measured, angle of dip.

(xi) Draw the geological boundaries on to just below the ground surface at their appropriate dip using a protractor.
(xii) You must now interpret the course of these boundaries both below the present ground surface and above it (the pre-erosional structure). This should be based on a combination of your field observations, your map and finding the simplest explanation for the geological structure.

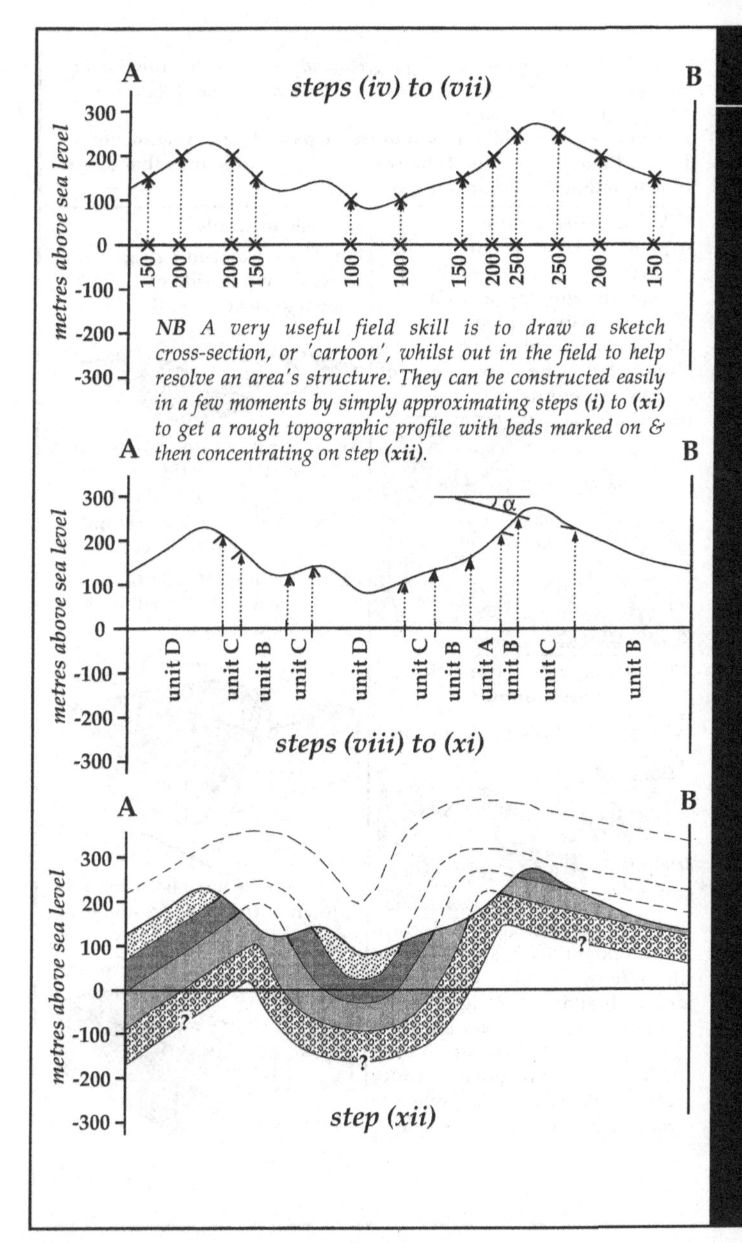
A
B
steps (iv) to (vii)
metres above sea level
300
200
100
0
-100
-200
-300
150
200
200
150
100
100
150
200
250
250
200
150
NB A very useful field skill is to draw a sketch cross-section, or 'cartoon', whilst out in the field to help resolve an area's structure. They can be constructed easily in a few moments by simply approximating steps (i) to (xi) to get a rough topographic profile with beds marked on & then concentrating on step (xii).
A
B
α
unit D
unit C
unit B
unit C
unit D
unit C
unit B
unit A
unit B
unit C
unit B
steps (viii) to (xi)
A
B
?
?
?
step (xii)

The whole purpose of constructing a geological map with an accompanying cross-section is that the two combined will allow the 3-dimensional structure of an area to be resolved.

Some exercises in this book require maps and cross-sections to be drawn. Below are some of the ways in which structures that might appear in them could be interpreted.

1. 'V-ing' across a valley

A planar, uniformly dipping bed or vein with a valley eroded through it, will appear to have a 'V' or 'U' shape in plan view, but this is simply an effect of the slope. **NB The direction of 'V-ing' depends on the dip direction.**

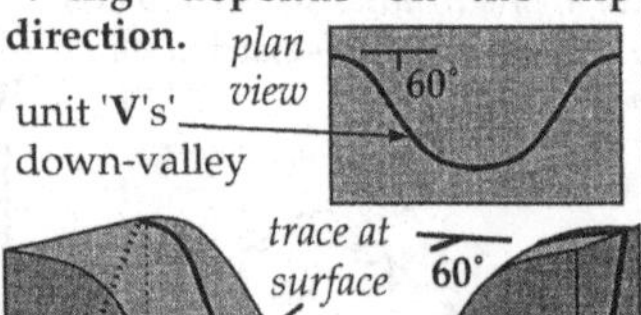

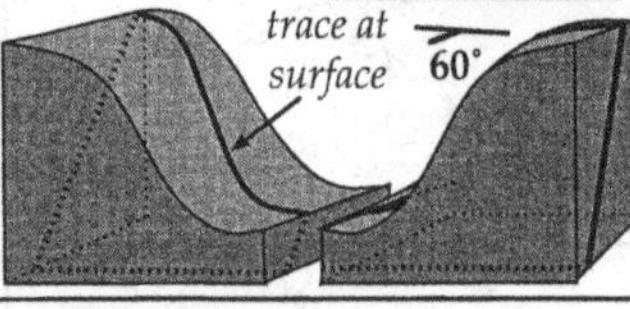

2. Vertical units & truncation at unconformities

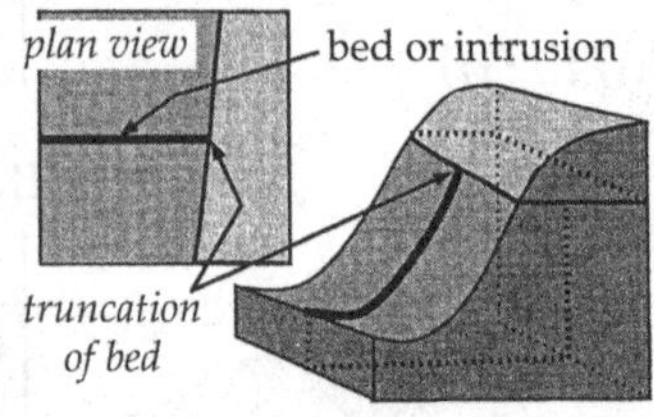

If a bed or intrusion is vertical then topography has no effect on its outcrop trace at the surface and it should be a straight line on a map. If it were overlain by a younger, nearly horizontal unit, then the **angular unconformity** between the two would appear to ***truncate*** the bed/intrusion below.

3. Repeating units

e.g. On a map, **unit A** appears to repeat either side of **B**. **A** is known to be younger than **B**.

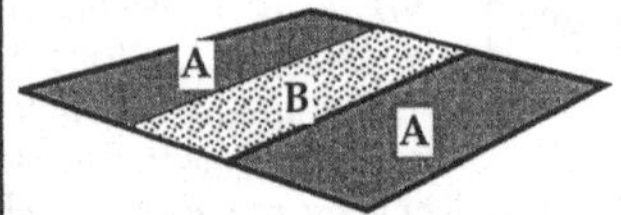

There are two different interpretations for this structure and a solution must be determined from the younging directions and a cross-section.

(i) A normal fault has dropped a block down & erosion has exposed the older **unit B**

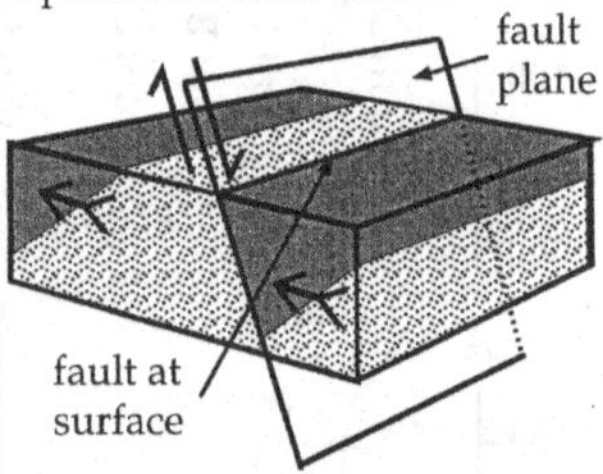

(ii) The units have been **folded into an anticline** & erosion has exposed **B** in the core

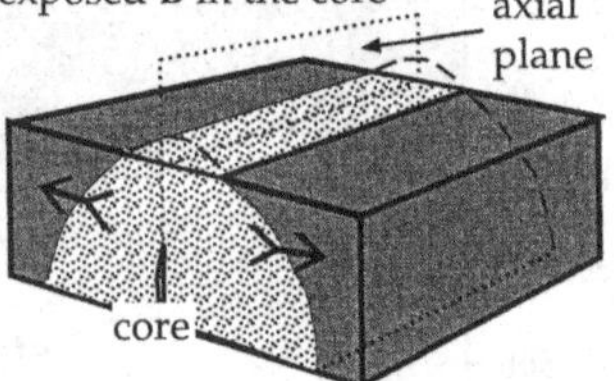

Further reading

If you would like to find out more about Arran's geological history or some of the other topics covered in the field exercises, then here are some books which you might find useful.

ARRAN

MacDonald, J. G. and Herriot, A. (eds.) 1983. *Macgregor's excursion guide to the geology of Arran*. 3rd edn., Geological Society of Glasgow, 210 pp.

McKerrow, W. S., and Atkins, F. B. 1989. *Isle of Arran: A field guide for students of geology*. 2nd edn., The Geologists' Association, 104 pp.

Tyrrell, G. W. 1928. *Memoirs of the Geological Survey of Great Britain: The geology of Arran*. 1987 Reprint, British Geological Survey, 296 pp.

Craig, G. Y. (ed.) 1991. *Geology of Scotland*. 3rd edn., The Geological Society, London, 612 pp.

McLellan, R. 1970. *The Isle of Arran*. Praeger Publishers, New York, 269 pp.

Hall, K. 1993. *North Arran – A postcard tour*. Richard Stenlake, Ayrshire, 56 pp.

DEVELOPMENT OF GEOLOGY AS A SCIENCE

Adams, F. D. 1938. *The birth and development of the geological sciences*. 1954 republication, Dover publications, Inc., New York, 506 pp.

Robson, D. A. 1986. *Pioneers of geology*. Special Publication of The Natural History Society of Northumbria, Newcastle Upon Tyne, 73 pp.

Hallam, A. 1989. *Great geological controversies*. 2nd edn., Oxford University Press, 244 pp.

Lyell, C. 1990. *Principles of geology; with a new introduction by Martin J. S. Rudwick*. 1st edn., University of Chicago Press, 3 volumes.

GENERAL

Leeder, M. R. 1982. *Sedimentology: Process and product*. George Allen and Unwin, London, 344 pp.

Hall, A. 1987. *Igneous petrology*. Longman Scientific and Technical, Longman Group UK Limited, 573 pp.

Park, R. G. 1983. *Foundations of structural geology*. Blackie and Son Limited, Glasgow, 135 pp.

Wyatt, A. (ed.) 1986. *Challinor's dictionary of geology*. 6th edn., University of Wales Press, Cardiff, 374 pp.

Further reading

Glossary

anticline – Folds in which the limbs dip away from the axis, closing upwards with the oldest rocks in the core of the fold. The strata form a 'Λ' in cross-section.

antiform – Fold in which the limbs dip away from the axis, closing upwards. The strata form a 'Λ' in cross-section.

baked margins – Any hot igneous body, whether intrusive or extrusive, will heat up and thermally alter any sediment which it comes into contact with. In particular, dykes and sills will do this to the country rock through which they intrude and when exposed at the surface these appear as *baked margins*.

bedding – Sediment is characteristically deposited in discrete 'packets' one on top of the other. This builds up a sequence of sedimentary *beds*, with the planar surfaces separating them known as *bedding planes*, *bedding surfaces*, or often just *bedding* for short.

bioturbation – A 'churning-up' of the sediment by organisms; typically these might be 'worms', but it can be caused by anything that burrows into the soft sediment and therefore disrupts the original depositional fabric.

chilled margin – The edges of a hot igneous body, whether intrusive or extrusive, will cool faster than the inside. More rapid cooling reduces the size of crystal that can grow before the temperature has dropped too far and so there is often a crystal size variation preserved in the igneous body once completely solid. The small crystals at the edge are recognised and described by referring to this region as the *chilled margin*.

clast – Any rock fragment that has been eroded, transported and deposited as part of a sedimentary bed is called a *clast*. Clasts can be of any size, so may be single mineral grains such as quartz in a sandstone, or reworked pieces of another lithology. A sedimentary rock composed of such clasts is called a *clastic sediment*.

competance – This is the inherent 'strength' and resistance of strata to deformation. The more competant or rigid a rock is, the less likely it is to undergo plastic deformation.

composite dykes – Where a series of igneous intrusions are injected along the same fracture in the country rock they can form a set of parallel, or *composite*, dykes.

convolute bedding – Sedimentary bedding can be disrupted prior to cementation and lithification. If the soft sediment has trapped a large amount of water within pore spaces it can become fluidised and the bedding may become 'deformed', or *convolute* in its appearance. This will also occur if the water is allowed to escape upwards to the surface.

country rock – A term applied to the surrounding rock through which an igneous intrusion or hydrothermal vein forces its way. *Country rock* can therefore be sedimentary, metamorphic or even igneous.

cross bedding – See [9]

desiccation cracks – A wet and muddy ground surface allowed to dry out will shrink and produce the polygonal cracks typical of desiccation.

devitrifying – An igneous glass can gradually equilibrate with the surrounding country rock over time and begin to crystallise. This growth of new crystals tends to take place around specific nucleation sites, producing a 'spotty' appearance to the rock. If this process continues to completion then the glass will change both colour and texture. The best example of this on Arran is the *devitrification* of the acidic glass, *pitchstone*, to crystalline *felsite* (see [5]).

dip – The angle between any structural feature and the horizontal is known as the *dip*. The angle of dip can be measured in a variety of planar surfaces such as bedding, cleavage, jointing or folding, but also the *dip* of lines on a surface can be found, such as the plunge of a fold axis (see [10]).

disconformity – If there is a sufficiently long break of deposition in a particular sedimentary environment, then this may lead to the development of a *disconformable* surface between the last bed to be deposited before the break, and the first to accummulate after sedimentation has resumed. There is no angular difference between the two beds but there may have been some erosion.

discordant – Dykes cut across any previous fabric in the rocks and so are said to be *discordant* with the country rock.

drift – Strictly speaking, any unconsolidated sediment between the soil and bedrock is called *drift*. However, only *glacial drift* is discussed in this book so the two terms are used synonymously.

dunes – see [9]

dykes – Sheet-like bodies of molten magma that intrude their way up through the Crust, cutting across the surrounding strata.

erratics – Glaciers or ice sheets carry an enormous amount of unsorted rock debris on, in and under them. When the ice melts this is all dumped on the bedrock. The larger 'boulders' that can often be seen strewn across the landscape of a glaciated area from this process are referred to as glacial *erratics.*

eustatic – This is a real change in global sea-level. This can be from melting or increasing the amount of polar ice, or by changing the volume of mid-ocean ridges (periods of increased rifting make the ridges 'hotter' and more buoyant, displacing ocean volume onto continental shelves and the sea-level appears to rise).

extrusive – Molten magma that manages to make it to the Earth's surface and escape forms *extrusive* igneous rocks such as volcanic lavas or ash (known as 'tuff').

fauna – A representative collection of the fossil organisms from a particular bed would be known as its *fauna*, or *faunal assemblage.*

fold plunge – See [18]

foliation – An alignment of minerals in a metamorphic rock can lead to it having a characteristic planar fabric or *foliation.*

geological contacts or boundaries – The surface of contact between two geological units (see [22]).

glaciers – 'Tongues' of ice that collect in cold and mountainous climates, and flow downslope within the confines of a valley (see **ice sheets**).

glass – Molten magma can be cooled so fast during intrusion that it does not have time to crystallise before the temperature has dropped too far. If this happens then the magma remains as a supercooled liquid, i.e. a glass. A superb example of this on Arran is the black igneous glass, *pitchstone.*

graded bed – A clastic sedimentary bed where the size of the clasts being brought in changed during its deposition. Typically, where a single 'pulse' of sediment settles out to form a bed, the larger grains are deposited first at the base and the grain–size decreases upwards until the finest clasts are removed from suspension. This leads to a 'fining upwards' in the bed and is a good **way-up structure** (see [9]).

groundmass – The micro-crystalline 'background' component of an igneous rock. Often, the composition of these crystals can only be approximated from the overall colour. This term is generally used where there are large crystals (**phenocrysts**) present and seemingly 'set' in this finer *groundmass.*

hydrothermal veining – 'Hot' waters percolating through rock at depth in the crust are termed 'hydrothermal fluids' and these can concentrate minerals such as silicates in solution. Veins can occur where this fluid is concentrated along a fracture followed by crystallisation. On Arran they are typically associated with tectonic deformation or igneous intrusions.

ice sheets – Thick accumulations of ice in mountainous and cold climates that flow away unrestricted by the confines of a valley (see **glaciers**).

intrusive - Molten magma that fails to make it to the Earth's surface and remains at depth in the Crust crystallises to form *intrusive* igneous rocks such as dykes, sills or plutons.

longshore drift – Prevailing tidal currents around coastlines may transport sediment laterally and this movement is termed *longshore drift.*

Ma (mega anni) – This is the abbreviation used by geologists for 'millions of years (ago)'.

magma mixing – See the background information section for Exercise 10.

massive beds – 'Packets' of sediment are deposited to form beds, but where these are particularly thick, homogeneous and perhaps resistant to weathering they are often described as being *massive.*

maturity – The progression of soft sediment during transport towards the final 'goal' of perfectly sorted sediment where all of the clasts have the same mineralogical composition, roundness, sphericity and grain size is called its state of *maturity.*

melt – A term used in conjunction with magma when describing the molten state of igneous rocks.

metasediments – Sedimentary rocks that have been buried to depth and metamorphosed, typically during episodes of tectonic deformation. This may result in the rocks developing a cleavage or the process may encourage the growth of new minerals to change the bulk composition of the rock (see [16]).

metamorphic cleavage – See [15]

non-sequence – Where a minor, short-lived break in deposition of sediment has occurred, the planar surface marking it is termed a *non-sequence.*

palaeogeography – 'Old geography'; literally the position of the World's continents or regions in the geological past.

phenocrysts – Crystals in an igneous rock that are clearly much larger than the main crystalline component, the **groundmass**, are called *phenocrysts.*

Phylum – A major division, or 'pigeon-hole' in zoological nomenclature. Invertebrate fossils on Arran fall into *Phyla* such as the Mollusca (bivalves, gastropods), Brachiopoda (brachiopods), Echinodermata (crinoids), Cnidaria (corals) (see [8]).

plutons/plutonic rocks – Plutons are large bodies of molten magma that have crystallised before reaching the surface. The northern granite of Arran is a good example of such a *plutonic rock*.

ripples – See [9]

shale – See [9]

sills – Sheet-like bodies of molten magma that intrude their way through the Crust by injecting between strata of the country rock. Therefore, they are *concordant* with sedimentary bedding.

stereonets – See [19]

stoping – Molten magma which partially melts and then incorporates blocks of the country rock into itself is said to *stope*. The term can also be applied in mining where pieces of a vein or country rock are broken off by the miners underground.

strata – Distinct layers of rock are termed *strata* and the study of their relationship to one another is *stratigraphy*. Abraham Werner was the first to establish this term and discipline (see Exercise 9).

strike – Originally coined by geologists from the German word '*to extend*' in order to describe the general compass direction in which outcrops were 'trending'; and it will still often be used loosely in this manner. Officially though, it now has a more strict definition. It is a horizontal line perpendicular to the dip direction of a bedding plane. Because it is a line, its value can be quoted as either one of two possible bearings 180° to each other, reflecting which way the compass was pointing on the line (see [10]).

syncline – Fold in which the limbs dip towards the axis, closing downwards with the youngest rocks in the core of the fold. The strata form a 'V' in cross-section.

synform – Fold in which the limbs dip towards the axis, closing downwards. The strata form a 'V' in cross-section.

transgression – Where the rock record demonstrates a relative rise in sea-level across a region, it can be termed a marine *transgression*.

trend – This is the compass bearing taken in the direction a particular feature points. Commonly *trends* will be measured for dykes or fold axial traces (see [14] and [18]).

twinning – Alkali feldspars commonly display the physical property of simple *twinning*. In hand specimen this appears as two euhedral crystals that have grown in unison but as perfect mirror images of each other.

Plagioclase feldspars will often grow as multiple twins, but typically this is on a much smaller scale and is best recognised in thin sections (see [5]).

unconformities – If there is a particularly long break in deposition at a locality then it may lead to the development of an unconformable surface between the last bed to be deposited before the break and the first to accummulate after sedimentation has resumed. During this time erosion will have left its mark and an angular difference between the two beds may result if there has been tectonic activity.

unit – A term applied loosely to group together a set of beds or strata that share the same characteristic physical properties. This is particularly useful when mapping an area.

veining – See **hydrothermal veining.**

way-up structures – Any sedimentary bed-forms, structures or fossils that will give an indication of which is the top of a bed and which is the bottom. In tectonically deformed areas like Arran, many beds are tilted and some are even vertical or overturned. Therefore, deciding whether or not overturning has occurred becomes critical in an area you are trying to map. All of the bed-forms in [9], except the first one, would be good *way-up structures.*

xenocrysts – **Phenocrysts** that are of a 'foreign' composition to the **groundmass** that they are found in are called *xenocrysts* (see [5]).

younging directions – This is the direction in which younger beds can be found in a sequence of strata (see [21], structural symbols). This is likely to be determined by finding *way-up structures* in the bedding.

Index by place name

Subject index
(see also Part 4)

Printed in the United States
By Bookmasters